IMMOBILIZED
MICROBIAL SYSTEMS:
Principles, Techniques
and Industrial Applications

Dr. Frieda B. Kolot started her career in industrial microbiology at a plant producing antibiotics. Six years later she began research in the field of caratinogenesis by fungi. Dr. Kolot was the author of several highly active strains of the single cell protein producer. After emigrating from the USSR to the USA, Dr. Kolot worked as a Georgetown University Fellow at the Institute of Allergies and Infectious Diseases (NIH) on interferon production by human and rabbit macrophages. Afterwards, at MIT, she investigated biotin production by *Candida utilis*. Dr. Kolot's interest in immobilized microbial systems resulted in solution of six difficult industrial projects, from which the series of her works in this field originated. Dr. Kolot is widely known in industrial microbiology by her numerous publications.

IMMOBILIZED
MICROBIAL SYSTEMS:
Principles, Techniques
and Industrial Applications

Frieda B. Kolot, Ph.D. [†]

ROBERT E. KRIEGER PUBLISHING COMPANY
MALABAR, FLORIDA
1988

Original Edition 1988

Printed and Published by
ROBERT E. KRIEGER PUBLISHING COMPANY, INC.
KRIEGER DRIVE
MALABAR, FLORIDA 32950

Printed in the United States of America.

Library of Congress Cataloging-in-Publication Data

Kolot, Frieda B.
    Immobilized microbial systems.

    Includes bibliographies and index.
    1. Immobilized cells. 2. Industrial microbiology.
3. Fermentation. I. Title.
 TP248.25.I55K65   1987   660′.62   86-34416
ISBN 0-89464-205-7

10 9 8 7 6 5 4 3 2

Contents

1

Immobilized Microbial Systems: Present State of Development

This introductory chapter familiarizes the reader with different techniques used for microbial attachment, their mechanisms, advantages, and limitations. The effects of carrier properties (composition, charge, surface area) as well as microbial properties upon immobilization are discussed. From the literature available, it seems that mainly adsorption, entrapment, and coupling were applied to microbial attachment. It is also clear that there is no universal technique and no ideal support. Only by properly matching cell and support properties can a stable microbe-carrier network be formed and used in continuously producing systems.

Introduction

Living microbial cells artificially bonded to insoluble supports have become a subject of increasing interest in the fermentative industry during the last few years. The advantage of such an approach compared with immobilized enzymes is obvious: Enzymes in the cell are maintained in their natural environment, neither extraction nor purification of the enzymes from microbial cells is necessary; thus, loss of enzyme activity is reduced. Preparation and utilization of immobilized microbial cells is easier than that of the immobilized enzymes, and enzyme stability is higher in whole cells. When cofactors are required, the use of microbes is preferred to enzymes since cells can regenerate the cofactors. Microbes can be attached without significant loss of catalytic activity; also, their operational and storage stabilities are high. This permits synthesis at large dilution rates in continuously producing systems, which usually need less capital investment and allow lower product cost because expensive fermentor design is eliminated. The purpose of this chapter is to acquaint the reader with the concept of "immobilized microbial systems" and to discuss their present state of development.

Methods of Attachment

The methods of microbial immobilization reported in literature have been categorized into four broad groups: 1. Immobilization without support 2. Adsorption 3. Covalent cross-linking 4. Entrapment.

1.0   Immobilization without Support

By contrast with immobilized enzyme systems, in which carriers are absolutely necessary for attachment, solid support is not always essential for microbial immobilization. The properties of microbial cells such as surface charge and cell wall composition are the major factors responsible for immobilization in such cases.

All *Aspergillus* strains can be easily immobilized without support (Metz, Kossen 1977). When spores of any of the *Aspergillus* strains are placed into continuous flow reactors, hyphae usually form agglomerates and become pellets. Hyphae-hyphae interactions are considered very important for microbial immobilization. The pellets are rather stable in airlift type fermentors and can be used for continuous production of citric acid or serve as supports for enzyme immobilization. (Hirano *et al.*, 1977)

Immobilization with Support

1.1   Adsorption

A great variety of different microorganisms such as species of *Bacillus, Pseudomonas, Saccharomyces* (Jack, Zajic, 1977), *Penicilliums*, and *Streptomyces* can adsorb to kieselguhr, wood, glass, ceramic, and plastic (Table 1). Some microbes attach because they form an adhesive disk to anchor the cell to the support. For others, a charge on the cell surface and cell wall composition provide the necessary electrostatic and ionic sites for attachment to the carrier. Affinity also depends on cell age. The amount of cells adhering varies from one genus to another. For the genus *Bacillus*, the proportion of cell attachment is small, but microbes which do adhere are very firmly held. By contrast, cells of

Table 1 Supports for Yeast Adsorption

Microorganism	Support	References
S. carlsbergensis	kieselguhr	Navarro *et al.*, 1976
	brick	Navarro *et al.*, 1976
	polyvinyl chloride	Navarro *et al.*, 1976
	wood	Durand and Navarro, 1978
	pouzzolane	Durand and Navarro, 1978
	dowex (acetate form)	Durand and Navarro, 1978
	doulite A 162	Durand and Navarro, 1978
	doulite A 101	Durand and Navarro, 1978
	doolase SP3	Durand and Navarro, 1978
	spherosil	Durand and Navarro, 1978
Candida tropicalis	ceramics	Marcipar *et al.*, 1979
Rhodotorula gracilis	ceramics	Marcipar *et al.*, 1979
S. cerevisiae	ceramics	Marcipar *et al.*, 1979
S. cerevisiae	Cordierite	U.S. Patent 4,149,937, 1979
	Spinel-zirconia	U.S. Patent 4,149,937, 1979
S. amurcae	fritted glass	U.S. Patent 4,149,937, 1979
	borosilicate glass	U.S. Patent 4,149,937, 1979

the genus *Pseudomonas* attach to the support in large numbers via a weak interaction. It would seem, therefore, that the adsorption phenomenon greatly depends on cell properties, but not totally upon them; carrier properties also affect microbe-carrier interactions.

The cell wall composition plays a large role in microbial attachment. Cells of *S. cerevisiae* and *Candida utilis* contain α-mannans, which have a strong affinity for cancanavalin A. Thus, after cancanavalin A is adsorbed to the chosen support, yeast cells will bind by affinity to the activated carrier (Horisberger, 1976). The mechanism of this technique is mainly electrovalent, but the strong affinity observed results from cell wall properties. In addition, one should remember that the complexity of the yeast cell wall surface can provide necessary amino or carboxy groups to directly interact with an inorganic carrier surface. This, too, results in formation of bonds between the cells and support.

Charge on the carrier surface is one of the major factors responsible for microbial attachment. For example, it is preferable to choose a positively charged support for *A. niger* immobilization, since the spores are negatively charged (as most yeast strains), (Sussman, 1966). The actual charges on glass and especially on ceramic surfaces are unknown, and this limits the choice for microbial attachment. On the other hand, pH of the solution in which immobilization is performed may modify the net potential on cell or on carrier surfaces and thereby greatly influence the charge-charge interaction (Chapter 2).

The advantage of using a porous support is well established for enzyme immobilization and is related to the amount of surface area available. Usually, it is also preferable to use a porous rather than a nonporous support for microbial attachment (Chapter 4).

The last factor which greatly influences microbial immobilization is carrier composition (Chapters 3, 4). The latter can provide metal ions which can react with amino or carboxy groups on the cell surface and promote linkages between the microbe and carrier.

The interactions leading to microbial attachment by adsorption evidently have a rather complex nature based on cell and support properties. By properly matching microbial and carrier characteristics, electrostatic interactions can form a stable cell-support network. Immobilization via adsorption has several advantages compared with other techniques: the cells remain alive and their enzymatic activity is not affected. The only disadvantage of this method is that electrostatic interactions are influenced by pH changes occurring during fungus metabolism.

1.2   Entrapment Techniques

1.2.1   Physical Entrapment into Agar (Jack, Zajic, 1977)

Cells of *Saccharomyces pastorianus* were immobilized by entrapment in a 2.5% (w/v) agar solution (50°C) with a volume ratio of cells to agar of 1:4. The mixture was quickly placed into a cold solution of toluene or tetrachloroethylene. Agar pellets with s200rical shapes were formed, in them the yeast cells were distributed homogeneously. Oxygen and product diffusion limitations are the major disadvantages of this technique.

1.2.2   Entrapment into Alginate Gel (Kierstan, Burke, 1977)

Alginic acid and its derivates are commercially available in a variety of types, with different gelling properties. Calcium alginate gels form rapidly in mild conditions and were found to be very suitable for immobilization. *Saccharomyces cerevisiae* cells were entrapped into a solution of sodium alginate. The suspension was slowly extruded into a 0.05 M $CaCl_2$ solution containing 10% (w/v) glucose, using sampler pipette with a 1mm diameter tip. Alginate gels provide a small barrier for diffusion of neutral substrate, but the method is extremely economical: only ten grams of support is required for immobilizing 200 grams of dry cells. Difficulties arise in the system when moderate levels of phosphate are present, which tend to disrupt the gel structure.

1.2.3 Entrapment into K-carrageenan Gel (Chibata, 1979)

K-carrageenan, a seaweed polysaccharide, is a non-toxic food additive which becomes a gel by cooling, as in the case of agar, or by contacting with an aqueous solution containing alkali, alkaline, bi- or trivalent metal ions, etc.

Suspensions of *Saccharomyces carlsbergensis* and K-carrageenan, dissolved in physiological saline, were separately warmed to 37–60°C, mixed, and cooled or contacted with an aqueous solution containing a gel inducing agent. The gel was granulated into particles with a suitable size and shape: cubic, bead, or membrane. To increase stability of the support, the matrix was treated with hardening agents such as glutaraldehyde or hexamethylenediamide. It was confirmed that yeast cells, immobilized by entrapment into K-carrageenan, were alive and able to multiply in the carrier, although some leakage of cells from the gel occurred during operation. The number of microbes in the matrix remained at the level of 10^9 cell/ml of gel. This method of immobilization did not affect cellular metabolic activity.

1.2.4 Entrapment into Polyacrylamide Gel (Ohmiya, et al., 1977; Yamamota, 1974)

Cells of *Saccharomyces lactis*, *S. fragilis*, *Candida pseudotropicalis var. lactosa*, and *Rhodotorula gracilis* were immobilized by entrapment into acrylamide gel. The polymerization solution consisted of cells—5–20%, acrylamide—15%, and cross-linking reagent—0.4–0.8%. Polymerization was initiated by adding tetramethylethelenediamide (TEMED) and accelerated by ammonium persulfate (APS). Cell exposure to free radical polymerization resulted in a decrease of enzymatic activity, which was mainly due to the toxic effect of acrylamide and TEMED. Thus, the concentrations of these components and the polymerization time had to be reduced. Several major factors influenced enzymatic activity of acrylamide entrapped cells: relative concentrations of cells, acrylamide monomer, TEMED, and APS, as well as the temperature of immobilization and the size of gel particles formed.

Microbes attached in this manner retained 61% of their initial activity.

1.2.5 Entrapment into Cellophane, Polyethylene, Caprone, and Phoroplast (Kaplackni & Isaeva, 1976)

Cells of *Saccharomyces carlsbergensis* were entrapped into four different materials. The rate of cell attachment was found to be a function of culture age, pH of the medium, and the presence of Ca^{2+} in solution. The highest rate of entrapment was observed during the first 15–30 minutes of contact. The supports immobilized from 55% to 65% of the cells, with small variations from one material to another.

1.2.6 Entrapment into Collagen (Venkatasubramanian, 1974)

This technique was developed by researchers at Rutgers University and is quite different from other methods of immobilization. It is based on the formation of non-covalent physico-chemical bonds such as salt linkages, hydrogen bonds, and Van der Waals interactions between carrier-protein and cell-protein. Cells (0.5–3%) are added to a collagen dispersion (pH = 6.5), and after initial flocculation, the pH of the aggregated mixture is slowly raised to 11.2. The dispersion is cast, dried at room temperature and tanned by dipping in glutaraldehyde solution. Chips of the membrane can be used in column packing. This method of immobilization results in the cooperative action of numerous relatively weak non-covalent bonds, which lead to the formation of a stable microbe-carrier complex.

1.2.7 Liquid Membrane Encapsulation (Jack, Zajic, 1977)

Cells of *Micrococcus denitrificants* were encapsulated in oil by adding the phosphate buffered cell suspension to a mixture of oil (86%), surfactant (2%), membrane strengthener (10%), anion transport facilitator, and then stirring at 600 RPM (18°C). The emulsion drops were 20–40 μm in diameter, each contained 500–600 cells. The emulsion thus formed was dispersed into a second phosphate buffered solution containing nitrate or nitrite substrate. Substrate reduction was demonstrated in batch and continuous studies. The system retained 78% of its activity after 5 days. Substrate/product diffusion through the liquid membrane were the limiting factors for this system.

1.3 Covalent Cross-linking

The mechanism of this method is based on covalent bond formation between activated inorganic support and cells, and requires the use of a binding (crosslinking) agent. To introduce the covalent linkage, chemical modification of the carrier surface is necessary. The advantage of this system compared with entrapment is that it is free from the diffusion limitation. Unfortunately, coupling agents are toxic and cells can retain only 60–75% of their enzymatic activity.

1.3.1 Silanization to Silica Support (Navarro, Durand, 1977)

S. cerevisiae microbes were immobilized by coupling to five types of silanized silica beads. Attachment consisted of two steps: introduction of reactive amino and carboxy groups onto the silica surface and reaction between activated support and yeast cells. γ-aminopropyltriethoxysilane was used as a coupling agent. This compound has dual functionality, resulting from an inorganic functional group at one end and an organic group at the other. The inorganic group condenses with a hydroxyl on the silica surface, placing the reactive organic group on the support surface. The second step involves covalent bond formation between organic groups on the silane treated carrier and suitable protein or carboxylate ligands from the cell surface.

1.3.2 Coupling by Glutaraldehyde (Navarro, Durand, 1977; Nelson, 1976) and Carbodiimide (Jack, 1977)

Cells of S. cerevisiae, S. aureus, B. subtilis, Ps. aeruginosa, and E. coli were immobilized by glutaraldehyde and carbodiimide coupling to silica beads. The mechanism of this reaction is identical to silanization: introduction of reactive organic groups onto the inorganic silica surface. The authors found that for non-activated silica, the number of cells retained on the carrier decreased with increasing surface area. When the support was activated, yeast cell attachment rose drastically with increasing surface area. These data indicate that larger support surfaces allow immobilization of more reactive organic groups, and therefore, more yeast cells can adsorb to the carrier. Without glutaraldehyde treatment, only a few groups on the silica could react with suitable ligands on the cell surface.

1.3.3 Coupling by Isocyanate

Cells of Saccharomyces cerevisiae and S. amurcae were coupled by the isocyanate coupling agent to different types of supports (fritted glass, borosilicate, zirconium-spinel, etc.) Retention capacities of the carriers declined after treatment, compared with simple adsorption.

1.3.4   Metal Hydroxide Precipitates (Barber et al., 1975; Kennedy et al., 1976)

This technique is based on a reaction between carboxyl and amino groups on the cell surface and the metal hydroxide. The result is a partial covalent binding of the microbial cells. Adding titanium or zirconium chloride salts to water results in the pH dependent formation of gelatinous metal hydroxide precipitates in which the metals are bridged to the hydroxyl or oxide groups. By conducting this reaction in a suspension of microbial cells, the microorganisms are entrapped in the gel-like precipitate formed. In such a gelatinous matrix, microbes are firmly held, and washings of the material are almost cell free. This method of immobilization does not decrease the metabolic activity of cells.

All glass and ceramic supports consist of varying proportions of aluminum, silicon, magnesium, zirconium, etc. oxides. Zirconia ceramic was found to be the best carrier for attachment of *Ps. chrysogenum* as well as *S. olivochromogenes*. This support incorporates zirconium ions into its structure. As a result of an ion exchange between the ceramic and the buffer solution in which immobilization is performed, zirconium hydroxide is formed on the carrier surface from zirconium oxide. The hydroxyl group of zirconium hydroxide can be replaced by a suitable ligand on the cell surface, resulting in a partial covalent linkage between the carrier and cell (Chapter 4).

1.3.5   Combinations of Coupling and Free Radical Polymerization (Nelson, 1976)

Various combinations of the methods described above can be used. Cells of *A. niger*, *Proteus rettgeri*, and *Rhodotorula gracilis* were first treated with glutaraldehyde and then exposed to free radical polymerization. Cells immobilized in such a manner retained fifty to seventy percent of their enzymatic activity.

1.4   Immobilization Techniques: Advantages and Limitations

Thus, there are three major techniques used for cell immobilization: adsorption, entrapment, and coupling. The mechanisms of these techniques are quite different (Table A). The basic reaction in adsorption is an electrostatic interaction between the charged support and charged cell. This technique is simple and takes no longer than thirty minutes. It should be emphasized that the charge on the cell surface changes with cell age and pH of the medium. The effect of surface charge on cell adsorption will be discussed in more detail in the second chapter.

Table A Immobilization Techniques: Their Mechanisms, Advantages, and Limitations

Immobilization Techniques		Factors Affecting Immobilization. Supports, Coupling Agents	Mechanisms	Advantages/Limitations
Adsorption (to organic or inorganic supports)	Cell properties	Cell wall composition Charge Age Dipolar character of the cell	Electrostatic interactions between carrier and cell surface	Simple/pH depended
	Carrier properties	Composition Surface charges Surface area pH Retention capacity		
Entrapment		Agar Alginate	Physical entrapment	/Substrate diffusion /Gel disruption by phosphates
		Pectate Collagen Carrageenan Plastic	Protein-protein interactions	
		Acrylamide	Free radical polymerization	Simple, low cost/Toxic
Coupling		Isocyanate Amino silane Glutaraldehyde Carbodiimide Metal hydroxide	Covalent bond formation Covalent bond formation Covalent bond formation Covalent bond formation Partial covalent bond formation	/Toxic Free from diff. limit /Toxic

Entrapment represents the most widely used technique. A great variety of different gels (agar, alginate, acrylamide, carrageenan, pectate, etc.) were tested as supports. In some cases, the only mechanism responsible for cell inclusion is entrapment (plastic materials). In other cases, in addition to physical entrapment, partial covalent bonds form between the cell and support, as is the case for pectate and carrageenan gels. If an acrylamide gel is used as support, the mechanism of cell attachment is free radical polymerization. This technique is severely damaging to the cells, and kills a large fraction of them. However, the enzymes can carry out the necessary reaction for rather long periods of time. By contrast, microbial cell entrapment into alginate does not harm the cells, but the gel itself can be disrupted if phosphate ions are present in the medium.

The coupling technique is quite different from adsorption and entrapment; it is based on covalent bond formation between activated support and cells. Each of the various coupling agents (aminosilane, isocyanate, carbodiimide, glutaraldehyde) places a specific group on the support surface, which later reacts with a reactive group on the cell surface. Unfortunately, coupling reagents can severely damage the cells.

1.5 Reactor Types

The types of reactors used for immobilized microbes are quite different from those in fermentation industry. They are water jacketed glass columns of various sizes, packed with microorganisms entrapped by various techniques. Aeration can be achieved by blowing sterile air through the columns, and the substrate can be continuously pumped from a reservoir. Because the reaction volumes of the columns are constant, they are often called fixed bed reactors.

Researchers from the Tanabe Seiyaku Co. in Japan developed a sectional packed column reactor. The difference is that the latter consists of ten sections, each with a 5.8 cm length and a 5.0 cm diameter, approximately 100 ml of volume prepared with a sampling pit. Each of the lower eight sections was packed with immobilized microbial cells. The substrate solution was preincubated in the top two sections to 37°C and then passed down through the column. The temperature in the column was maintained by circulating water in the outer jacket. This type of reactor allowed estimating the percent of conversion of substrate to product in the outlet of each section. The rate of reaction in packed columns of immobilized microbes is influenced by particle size, intra-particle concentration of cells, and external levels of substrate (Toda, 1975; Toda, Shoda, 1975). Other reactor types such as the fluidized tapered bed (Scott and

Hancher, 1976), and hollow fiber membrane systems (Kan and Shyler, 1978) were also tested for product formation by immobilized microbes.

1.6 Summary

The potential of immobilized microbial technology is just being recognized. A successful catalyst greatly depends on properly choosing a technqiue, a support, and a strain for attachment, and should minimize any adverse effect on the enzymatic activity of cells. The mechanisms of attachment are electrostatic interaction between charged cells and support, simple entrapment, covalent bond formation, and free radical polymerization with covalent bonding. Although some techniques (entrapment into acrylamide and coupling) are rather toxic and reduce the enzymatic activity of yeast cells, others seem to provide very mild conditions for microbe immobilization. In a few cases, cells are not affected at all, they even multiply inside the support because it does not present a barrier for product or substrate diffusion.

Evidently, there is no ideal or universal technique and no ideal support. What seems to be good for one strain and carrier can not be simply transferred to others. Due to the disadvantages of entrapment into acrylamide and coupling, researchers are now turning toward adsorption onto different supports and entrapment into gels other than polyacrylamide.

Adsorption depends on cell age, retention capacity of the carrier, pH, and electrostatic interactions between the cell and support. Entrapment seems to have more advantages because it is simple and a high microbe concentration can be achieved. Only a limited number of publications have been devoted to a relative comparison of the effectiveness of different techniques, all of which (except adsorption and entrapment in metal hydroxide precipitates) lower cell enzymatic activity. Since 1973, six industrial processes have been commercialized, four in Japan under the direction of I. Chibata (Tanabe Seiyaku Co.) and two in Europe (Tosa et al., 1973, 1974; Yamamota et al., 1974, 1976, 1977; Sato et al., 1976). While Japanese researchers continue concentrating on one step enzymatic reactions and are utilizing gel entrapped cells, European investigators immobilized cells by adsorption. One should keep in mind that the initial choice of support and method for immobilization can determine success or failure of an immobilized microbe application program.

Literature

1. Dainty, A. L. et al., Biotech. Bioeng., **28**, 12, 210 (1986)
2. Schnaar, R. L., Langer, B. G., Brandley, B. K., Anal. Biochem., **151**, 2, 268 (1985)
3. Chang, H. N., Park, T. H., J. Theor. Biol., Sept., 116, 9 (1985)

4. Powell, L. W., *Biotechnol. Genet. Eng. Rev.,* **1**, 1 (1984)
5. Mauz, O., Noetzel, S., Sauber, K., *Ann. NY Acad. Sci.* **434**, 251 (1984)
6. DeLuca, M., *Curr. Top. Cell. Regul.,* **24**, 189 (1984)
7. Varani, J., *Biotech. Bioeng.,* **25**, 5, 1359 (1983)
8. Hahn-Hagerdale, B. *et al.*, *J. Chem. Tech. Biotech.,* **32**, 15, 7 (1982)
9. Thonart, P. *et al.*, *Enzyme and Microbe Tech.,* **4**, 191 (1982)
10. Kolot, F. B., *Proc. Biochem.,* **16**, 6, 30 (1981)
11. Kolot, F. B., *Proc. Biochem.,* **16**, 5, 2 (1981)
12. Messing, R. A., Oppermann, R. O., Kolot, F. B., *Biotech. Bioeng.* **19**, 59 (1979)
13. Ohlson, S., Larson, P. O., Mosbach, K., *Biotech. Bioeng.,* **20**, 1267 (1978)
14. Kan, J., Shuler, M., *Biotech. Bioeng.,* **20**, 217 (1978)
15. Jack, T. R., *Biotech. Bioeng.,* **19**, 631 (1977)
16. Hirano, K., Karube, I., Matsunaga, T., Suzuki, S., *J. Ferm. Tech.,* **55**, 4, 401 (1977)
17. Metz, B., Kossen, N., *Biotech. Bioeng.,* **19**, 781, (1977)
18. Yamamota, K., Tosa, T., Yamashita, K., Chibata, I., *Biotech. Bioeng.,* **19**, 1101 (1977)
19. Yamamota, K., Tosa, T., Yamashita, K., Chibata, I., *Eur. J. Appl. Micro.,* **3**, 169 (1976)
20. Sato, T., Tosa, T., Chibata, I., *Eur. J. Appl. Micro.,* **2**, 153 (1976)
21. Barber, S. A., Kay, I. M., Kennedy, J. F., U.S. Patent 3,912,593
22. Jack, T. R., Zajic, J. E., *Advances in Biochemical Eng.,* **5**, 126 (1977)
23. Kennedy, J. F., Barber, S. A., Humphreys, J. D., *Nature,* **261**, 242 (1976)
24. Larson, P. O., Ohlson, S., Mosbach, K., *Nature,* **263**, 796 (1976)
25. Martin, C. K. A., Perlman, D., *Biotech. Bioeng.,* **18**, 217 (1976)
26. Nelson, R. P., U.S. Patent 3,957,580
27. Scott, C. D., Hancher, R. C., *Biotech. Bioeng.,* **18**, 1393 (1976)
28. Shimizu, S., Marioke, H., Tani, Y., Ogata K., *J. Ferm. Technology.,* **53**, 2, 77 (1975)
29. Slowinski, W., Charm, S. E., *Biotech. Bioeng.,* **15**, 973 (1973)
30. Sussman, A., *The Fungi.* Academic Press (1966)
31. Tosa, T., Sato, T., Mori, T., Maruo, Y., Chibata, I., *Biotech. Bioeng.,* **15**, 69 (1973)
32. Tosa, T., Sato, T., Mori, T., Chibata, I., *Applied Microbiology,* **27**, 5, 886 (1974)
33. Toda, K., *Biotech. Bioeng.,* **17**, 1729 (1975)
34. Toda, K., Shoda, M., *Biotech. Bioeng.,* **17**, 481 (1975)
35. Venkatasubramanian, K., Saini, R., Vieth, W. R., *J. Ferm. Technology,* **52**, 4, 268 (1974)
36. Yamamota, K., Sato, T., Tosa, T., Chibata, I., *Biotech. Bioeng.,* **16**, 1589 (1974)
37. Yamamota, K., Sato, T., Tosa, T., Chibata, I., *Biotech. Bioeng.,* **16** 1601 (1974)

2

Changes in Cell Properties
after Immobilization

In this chapter changes in cell properties after immobilization will be discussed. These changes will be influenced by the cell wall composition, method of immobilization, matrix type, and cell type. We will focus on the effects of immobilization upon cell concentration and physiological stages, changes in cellular constituents, shifts in pH and temperature optima, operational and heat stabilities, disappearance/shortening of lag time, etc.

2.0 Changes in Yeast Properties after Entrapment into K-carrageenan Gel

Experiments with free and immobilized yeast cells were started with equal amounts of seed culture (3.5×10^6 cells/ml of gel). Both cultures were grown for 60 hours under similar conditions in a complete medium. Total cell counts were measured at equal time intervals, revealing that the number of cells in both systems increased linearly between ten and forty hours of incubation, and reached 4×10^8 cells/ml after that period. Free cells reached a steady state of 2×10^8 cells/ml after 60 hours of incubation, while immobilized cells simultaneously reach steady state at the much higher concentration of 2×10^9 cells/ml of gel. The one log difference in cell number was confirmed by calculation of cell weight and microscopic observation.

Microscopic observation of entrapped cells also showed that immobilized microbes grew inside the matrix, forming a cell layer near the gel surface, where nutrients were most easily available. Diffusion limitations of the nutrients prevented cells from growing in the central core of the carrageenan gel beads. Although the three hour generation time of immobilized yeasts in the exponential growth phase was similar to that of free cells, product formation of the immobilized cell system during steady state (after 60 hours of incubation) was clearly superior to that of the free cell system, which was attributed to a higher average concentration of gel entrapped cells compared with freely suspended cells.

Navarro and Durand[25†] also noticed several changes in *Saccharomyces carlsbergensis* and *S. cerevisiae* yeast physiology after immobilization. Substrate consumption as well as alcohol and CO_2 production of the two systems were compared during fermentation. It was established that immobilized yeasts showed a six fold increase in specific alcohol production, a five fold increase in specific substrate consumption, and a five fold increase in specific CO_2 production. (Glucose oxidase enzyme activity of *E. coli* adsorbed to a filter membrane was also higher (26%) than that of free cells.[31,32†])

These findings were later confirmed.[5†]. Working with the *Saccharomyces cerevisiae* strain, Marcipar *et al.* found that the respiration rate of fixed cells was higher compared with free cells. A five fold greater oxygen intake by immobilized *S. cerevisiae* microbes was also recently noted by Vijaya-lakshmi[2†]).

Unfortunately, the changes in cell metabolism are not always favorable to the microorganism. A decrease in oxygen consumption for *Candida lipolitica* was reported and attributed to toxicity of the carrier.

These data thus confirm that cell metabolism changes after immobilization due to support contribution to the biocatalyst system. Several hypotheses can be offered to explain this phenomenon. The first proposes that upon contact with a cell, a carrier can activate the membrane effector governing cell respiration. A second explanation could be that carriers modify membrane permeability, which may result in greater availability of nutrients and O_2. Another possibility is that carriers can modify the environment surrounding the cells. Whatever the exact mechanism of cell activation in the immobilized state is, clearly, the support surface does contribute to microbial activity.

Changes in metabolism due to immobilization were noticed mainly for bacterial and yeast cells; not surprisingly, processes based on immobilized bacteria and yeasts have already been industrialized.[7,8†]

---

†Refer to literature of Chapter 3.

2.1 Changes in Yeast Properties after Entrapment into Acrylamide Gel

2.1.1 Effects of Immobilization on Physiological Stages of Yeast

The acrylamide monomer is toxic to all cells and requires special precautions when it is used as a support material. Most recent data revealed that the gel is especially harmful to young cells in the exponential growth phase (15hr) and to cells at the beginning of the stationary phase (24hr), compared to cells in the stationary phase (48hr). Equal amounts of cells at various stages of growth (15, 24, 48hr) were immobilized by entrapment into acrylamide gel, and the percentage of microorganisms which survived polymerization was estimated. Fifty-five percent of cells in the stationary phase survived polymerization, while the percentage of survivals in the exponential growth phase and the beginning of the stationary phase reached only half of that level (22% and 28%, respectively).

Therefore, when using the acrylamide support for immobilization, an increase in cell count by a factor of ten is recommended to compensate for the extreme sensitivity of young cells to monomer toxicity. Only in this case can one expect to obtain at least 45% survival inside the matrix.

2.1.2 Changes in Macromolecular Constituents

A comparison of the macromolecular constituents of immobilized and free cells showed that the protein: macromolecular component ratios differed for free and fixed microbes during their growth cycles. After forty-eight hours, immobilized cells contained more DNA, glycogen and glucan, but less trehalose, which exhausted more rapidly than other reserves. Depletion of this carbohydrate occurred because entrapped microbes operated under substrate diffusion limitations and utilized their own reserves.

Observed changes in cell wall constituents (mannan and glucan) may be a result of the acrylamide gel effect during polymerization.

2.1.3 Cell Integrity after Entrapment into Acrylamide Gel

Cell entrapment into acrylamide resulted in loss of viability in a large portion of the population, with destruction occurring in all cell types used, but varying in extent from one strain to another. These variations may well have resulted from differences in cell wall composition. More than 99% of Gram negative bacteria, *Paracoccus denitrificants*, were destroyed, while the Gram positive bacteria, *Lactobacillus casei*, were only 30% destroyed. The toxic effect of acrylamide gel on yeasts could be partially avoided if cells were immobilized in the stationary phase, as already mentioned.

Yeasts entrapped into the matrix were subdivided into cells which were fully destroyed (75–85%) and survivals (25–15%). Survivals were further subdivided into a) normal cells with scars and b) living cells without scars. The latter often formed pseudomycelia. The cell subdivision was confirmed in experiments with fluorocurom calcofluor S.T., followed by electron microscopy. The reagent specifically stains the bud scars of intact cells because it is fixed by cell wall polysaccharides, while uniformly staining the entire wall surfaces of destroyed cells. Thus by applying this stain, not only cell age but also cell integrity can be studied.

Electron microscopy of cells reisolated from the acrylamide matrix showed that they were mostly lysed to varying extents, although cells with normal structure could be found. Cell walls of entrapped cells appeared to be much thicker and composed of two layers. The second layer, more electron dense than the first, consisted of acrylamide gel directly bound to the wall surface.

Several other changes which occurred after immobilization and their possible mechanisms will be briefly summarized. Detailed discussions of specific cases are presented in separate sections.

2.1.4   Shifts in pH Optima

Generally, pH optima shifted 0.5–1.0 units at peak enzymatic activity of immobilized *Nitrobacter agilis*, *Brevibacterium flavum* (malic acid), *Pseudomonas docunhae* (L-alanine), *E. coli*, and *Azotobacter agile*. Immobilized microbes also exhibited broader pH optima, which suggests a matrix contribution to the catalyst system and emphasizes the importance of support composition. The cationic layer of inorganic matrices may activate enzymes located on the cell surface, whereas the metallic ions added for hardening and stabilizing the matrix of organic supports may undergo oxidation-reduction and thus activate enzymes located on the cell membrane.

2.1.5   Shifts in Temperature Optima

Increased temperature optima (compared with free cells) were noted for L-alanine production via *Pseudomonas docunhae* cells attached to K-carrageenan matrix, for L-malic acid production via *Brevibacterium flavum* cells also immobilized into K-carrageenan gel, and for *Pseudomonas pubida* cells immobilized into an acrylamide matrix. In some cases, processes were optimally performed at temperatures nearly 10°C higher compared to free cells, thus lessening the chance of contamination.

2.1.6 Changes in Operational Stability

A major advantage of the microbial catalyst lies in its long term operational stability compared to the batch-wise free cell process and the immobilized enzyme process.

The half-life of a *Brevibacterium flavum* catalyst used for malic acid production was 160 days. The microbial catalyst formed by entrapment of *E. coli* ATCC 11303 cells into K-carrageenan gel (L-aspartic acid production) had a half-life of 680 days. An *S. cerevisiae* catalyst used for continuous ethanol production exhibited a half-life of 80 days, and the half-life of the immobilized cell system used for side-chain cleavage at the C_{17} position was more than 40 days.

One explanation for the excellent operational stability of immobilized cells compared to immobilized enzymes is the elimination of enzyme leakage and enzyme inactivation. This is a direct result of immobilized cells living and multiplying inside the support matrix, where they regenerate cofactors and perform single step as well as multi-step enzymatic reactions.

2.1.7 Shortening or Disappearance of Lag Time

Free *E. coli* and *Nitrobacter agilis* cells used for succinate oxidation exhibited a lag time of 1.5–2.0 hours in oxygen uptake, while cells adsorbed to Dowex resins and used for the same purpose exhibited no lag time. Immobilized *E. coli* used for lactose oxidation had a much shorter lag time than free cells. These results suggest a carrier activation of cell membrane enzymes.

2.1.8 Strength of Cell Fixation to Support

To determine the strength of cell fixation, yeasts were first adsorbed by recirculation and then washed (at a flow velocity not exceeding 2.0 cm/min) in order to remove the non-adsorbed cells. The number of washed-off microbes was then recorded vs. flow speed, as the latter increased from 2.3 cm/min to 10.0 cm/min. The fluid velocity causing an appreciable rise in cell concentration of the medium was considered the minimum flow capable of creating hydrodynamic forces sufficient to detach the cells. For *Saccharomyces cerevisiae*, in-

Table B   Changes in Cell Metabolism After Immobilization

1. The total number and concentration of cells in the matrix increases. It is not uncommon to observe a tenfold rise.
2. Microbe generation time is usually shorter.
3. Changes occur in macromolecular cell components (more DNA, glycogen, and glucan but less trehalose).
4. pH optimum shifts 0.5–1.0 units.
5. Temperature optimum shifts about 10°C.
6. Operational stability usually increases after successful immobilization (*Brevibacterium flavum* for malic acid production—160 days; *E. coli* ATCC 11303 for aspartic acid production—620 days).
7. In some cases, lag time disappears.
8. Respiration rate, oxygen consumption, enzymatic activity increase.

creasing pH of the medium from 4.0 to 6.0 reduced the minimum flow rate from 4.7 cm/min to 3.4 cm/min. For *Candida tropicalis* cells, raising pH of the medium by the same amount increased the minimum flow rate from 3.5 cm/min to 4.7 cm/min. The data led the authors to conclude that strength of cell fixation to a support is characteristic of the strain and is independent of pH.

2.1.9   Effect of Cell Age on Adsorption

To determine a correlation between cell age and adsorption, the mean cell volume before immobilization ($36.25\mu^3$) and the mean cell volume of unadsorbed cells after immobilization ($27.5\mu^3$) were estimated. Since older cells are larger than younger cells, it was concluded that the older cells were adsorbed more so than their counterparts. It therefore seems logical that immobilizing older microbes would prove more profitable in the long run because the greater number of cells adsorbed per unit carrier would carry out the needed reaction more quickly.

Literature

1. Doran, P., Bailey, J., *Biotech. Bioeng.*, **28**, 1, 73 (1986)
2. Dainty, A. *et al.*, *Biotech., Bioeng.*, **28**, 2, 210 (1986)
3. Pastorino, A., *J. Appl. Biochem.*, **7**, 2, 122 (1985)
4. Zueva, N., *Prikl. Biokhim.*, **21**, 3, 333 (1985)
5. Chang, H., Park, T., *J. Theor. Biol.*, Sept, 116, 9 (1985)
6. Adlercreutz, P. *et al.*, *Applied Micro. Biotech.*, **20**, 5, 296 (1984)
7. Kumakura, M., *Biosci Rep.*, **4**, 3, 181 (1984)
8. Minard, P., Legoy, M., *Ann. NY Acad. Sci.*, **434**, 259 (1984)

9. Thonart, P. *et al.*, *Enz. Micr. Tech.*, **4**, 191 (1982)
10. Siess, M., Davies, C., *Eur. J. Appl. Micro. & Biotech.*, **12**, 10 (1981)
11. Wada, M. *et al.*, *Eur. J. Appl. Micro. & Biotech.*, **10**, 275 (1980)
12. Wada, M. *et al.*, *J. Ferm. Tech.*, **58**, 4, 327, (1980)
13. Navarro, J., Durand, G., *Eur. J. Appl. Micro. & Biotech.*, **4**, 243, (1977)

3

Organic Supports
for Microbial Attachment

The purpose of this chapter is to discuss the most important support criteria, strategies for carrier selection, and mechanisms of microbe-support interactions. Properties of carbon-based supports are presented: reactive groups on the carrier surfaces, retention capacities, as well as roles of support composition, charge, surface area, mechanical strength, particle shape and size, etc. Attention is focused on carriers which provide: 1. high retention capacity 2. no decrease in ezymatic activity 3. satisfactory performance during column operation. Changes in cell metabolism due to carrier contributions, their possible mechanisms, and strength of cell fixation to the support are presented.

Selection of inorganic carriers will be discussed in Chapter 4.

Introduction

The application of living immobilized microbial cells as biocatalysts represents a new, fascinating and rapidly growing trend in microbial technology. By contrast with presently accepted batch or continuous fermentations, in which free cells are utilized, immobilized cells exhibit many advantages: the reaction rates can be accelerated due to increased cell density per unit reactor, wash out seldom occurs even at high dilution rates, cell metabolism and cell wall permeability increase due to contact with support, cells can multiply inside the support and, if needed, can be activated in the immobilized state, costly steel fermentors, usual nowadays at fermentation plants, could be eliminated.

Instead, modern plants would be smaller, comprised of columns packed with immobilized cells. As a result, better process control could be achieved.

Microbial biocatalysts offer several advantages compared with immobilized enzyme systems since costly enzyme extraction, isolation, and purification are obviated. The enzymes are also more stable inside the cell due to their localization in a natural environment, and there is no need for cofactor regeneration.

However, successful microbial cell immobilization and long term stability of biocatalyst during continuous operation greatly depend on choosing the proper support as well as the proper method for cell attachment.

3.0   General Considerations

Supports used for microbial attachment can be divided into two main categories: organic and inorganic. Although much literature is devoted to showing the superiority of each type, it is widely accepted that organic carriers are better, since they provide a larger variety of reactive groups on their surfaces, such as carboxyl, amino, hydroxyl, etc. The organic supports can be subdivided into carrier polysaccharides, proteins, and synthetic polymers. Depending on the subunit, polysaccharide supports can be further subdivided into cellulose and its derivatives, dextrans or agaroses, while synthetic polymers can be subdivided into acrylamide polymers, copolymers of phenol and formaldehyde (phenolic resins), copolymers of styrene and divinylbenzene (polystyrene polymers), and urethane polymers (Table 1). Inorganic materials which comprise various oxides and their combinations, exhibit high thermostability and good flow properties. In solution their surfaces are hydrolyzed, and the hydroxyl groups formed can directly react with carboxyl or amino groups on the cell surface. This reaction usually occurs on the surface of all known inorganic, so-called ungrafted supports. However, besides ungrafted, there are grafted inorganics, e.g. inorganic carriers to whose surfaces special organic groups are attached.

Since the microbial cell wall, although quite different for bacterial cells, yeasts, and fungi, is comprised of a variety of proteins (polysaccharides [mannans, theichoic acids], amino sugars, etc.) the mechanism of interaction between the microbial cell and support can be reduced to electrostatic interactions between charged cell and charged support; also, ionic, partial covalent, and covalent bonds form between the support and cell.

Table 1 Microbial Support Classification

Organic	Inorganic
Polysaccharides	Ungrafted
Carrageenan	Alumina
Cellulose	Zirconia
Agar	Magnesia
Pectate	Silica
Agarose	Glass
Sephadex	Ceramics
Protein	Ferromagnet
Collagen	Grafted with various coupling agents
Synthetic polymers	
Acrylamide	
Phenolic resins	
Polystyrene polymers	
Urethane prepolymers	

3.1 Support Criteria

Only brief and general consideration of support criteria which need to be kept in mind when choosing a carrier for microbial attachment will be stressed here. Later in the text, the advantages and limitations of each specific carrier will be discussed. One of the most important support criteria is that it should be nontoxic to the cells and should not affect their metabolism. Ideally, one can find a support which increases cell membrane permeability or enhances cell metabolism. However, the minimum we are looking for is coexistence between the microbe and the support. A second support criterion is its retention capacity. It is absolutely necessary that the support retain a high microbial loading, which is defined as the amount of cells (dry weight protein) incorporated per one gram of carrier. The retention capacity of a support can be found as the difference between the amount of cells initially added to the support and the amount which did not react with the support, and were therefore washed out. Obviously, for high loading it is preferable to choose entrapment into polysaccharide or collagen type gels vs. adsorption/coupling to inorganic supports (Table 2). Here, it is necessary to note the difference between enzyme and immobilized cell preparations in terms of support loading. It is established that with a high

Table 2   The Retention Capacities of Various Supports

| Support | Amount of Cells Attached, Expressed as | | | References |
	Percent of Initial Amount	Number of Cells per ml. Support or mg. Cell per gr. Support	Protein (γ) per Gram Support	
1. Polysaccharides				
Carrageenan	100	10^9 cell/ml.	—	1
Pectate	50	—	—	2
2. Protein				
Collagen	50	—	—	3
3. Synthetic polymers				
A. Dualite A-162		9 mg./gr.		4
B. Dualite A-101		17 mg./gr.		4
C. Dowex		21 mg./gr.		4
D. PVC		80 mg./gr.		4
4. Inorganic supports				
A. Biodamine ceramic	40–70			5
B. Fritted glass	—	—	11.3×10^3 γ/gr.	6
C. Cordierite ceramic	—	—	12.8×10^3 γ/gr.	6
D. Zirconia ceramic	—	—	13.4×10^3 γ/gr.	6

increase in enzyme loading, immobilized enzyme preparation activity decreases due to protein-protein interactions. By contrast, an increase in microbial loading from 10^6 to 10^9 cells per milliliter of gel does not result in any decline of cell activity. Knowing the exact charge on the support surface as well as its composition or presence of specific reactive groups, provide the basic information on the types of interactions and bonds which may be expected to form between the cell and support. The carrier should have the capacity for forming a large number of such bonds.

Due to their application in biotechnology, where sterility of the entire process is absolutely essential, the supports should be stable at elevated temperatures. In addition to thermal stability, a carrier should be stable in the specific pH range of the process it is performing. Carriers with high porosity are desirable since they allow large microbial loading as well as good substrate diffusion to the cell. A support should be stable under high pressure and should have specific particle shape and size. This makes reactor operation easy and may permit a high substrate flow rate.

It is necessary that the support material be relatively inexpensive and preferably with carrier regeneration. However, the latter requirement is not always necessary.

3.1.1   Carrier Thermostability

The thermal stability of a support plays an important role in carrier evaluation because microbial cells can only be immobilized to sterile carriers, which requires that the latter be stable at elevated temperatures. If a support is not thermostable or if its properties change with exposure to high temperatures, it should be cold sterilized through various membrane filters. Sterilization by U.V. and chemical methods should be avoided due to risk of reactions changing support surface characteristics.

3.1.2   Carrier Shape

The shape of a carrier affects the performance of an immobilized cell system in a continuous flow reactor. Different support geometries: fibrous, membrane, bead, and cubic are available or can be prepared. Detailed methods of forming porous bead particles from any type of cellulose and particles of various shapes from carrageenan gel can be found.[10,12] Only fibrous carriers should be avoided in column operation due to poor flow properties.

The stability of a particle shape is very important. It was shown that carrageenan gel beads are rather stable and can keep their shape for at least three months of operation. Vieth and Venkatasubramanian stated[3] that membranous collagen is also an excellent support for various microbial immobilizations. When rolled into a spiral and placed inside a column, it maximizes contact between the cells and substrate and produces a very low pressure drop even when operating with substrate matter which would cause plugging problems in conventional fixed bed reactors. Another advantage of this type of support is the potential for inserting a large amount of membrane into a confined reactor volume, thereby increasing the reaction rate.

3.1.3   Carrier Particle Size and Mechanical Strength

Carrier size and mechanical strength affect process performance in a reactor. Larger carrier particles cause smaller pressure drops in a column and allow higher substrate flow rates. For inorganic supports, it was found that particles with a size greater than or equal to 50 mesh perform well during column operation.

Strength of the support is another important characteristic of the carrier. Vijayalakshmi *et al.*[2] stated that cross linked polysaccharide pectate with 500 μmoles of Fe^{+3} per gram dry gel could withstand pressures of more than 2000 psi and exhibited very good flow properties up to 2000 ml/cm²/hr.

3.2   Mechanisms of Microbe-support Interactions[7]

There are five different mechanisms responsible for microbial attachment to any support (Table A, Chapter 1): 1. electrostatic interactions between charged cells and charged support 2. ionic bond formation between amino NH_3^+ and carboxyl COO^- groups on the cell surface and a reactive ligand on the support surface 3. partial covalent bond formation between amino (carboxyl) groups on the cell surface and hydroxyl (or dissociated silinol) groups on the support surface, as in the case of glass and ceramics 4. physical entrapment (plastics), where pores of the support are smaller than the microbial cells; thus, cells remain inside the carrier, while nutrients can move in and products can move out 5. covalent bond formation between reactive groups on the cell surface and specially grafted groups on the support. In the last case the carrier is activated and spacer arm is inserted between the cell and support. The reactive groups on carrier surfaces most often activated by coupling agents are hydroxyl groups, present on organic (cellulose, dextran, synthetic polymers) as well as on inorganic supports, and carboxylic groups, which can be converted into esters and can later react directly with amino groups, thus creating amide bonds.

The first three mechanisms appear simultaneously and are common for all supports, but for covalent bonds to form, special coupling agents (isocyanate, amino silan, glutaraldehyde, carbodiimide, etc.) are necessary. Although the binding mechanisms responsible for attachment of microbial cells to supports are complex, to achieve maximum microbial loading, it is absolutely necessary to evaluate microbial as well as carrier properties in great detail, keeping in mind that microbial cells should remain alive and their properties unaffected.

Bacterial cells, yeasts, and spores of the genus *Aspergillus* are negatively charged. Therefore, if adsorption is the technique of choice, it is preferable to choose positively charged supports such as DEAE Sephadex A-50, DEAE Sephadex A-25, Amberlite IR-45, or Bio-Rad AG-21 K, than negatively charged inorganics or ceramic supports.

Table C   Factors Affecting Immobilization

Support Criteria
 1. High retention capacity
 2. Nontoxic
 3. Specific size and shape
 4. High mechanical strength
 5. Preferably a porous material
 6. Thermostable
 7. Stable at pH of the process
 8. Reusable

Mechanisms of Cell-Support Interactions
 1. Electrostatic interactions
 2. Ionic bond formation
 3. Partial covalent bond formation
 4. Physical entrapment
 5. Covalent bond formation

3.3   Organic Supports

With all this information in mind, and recalling the methods of immobilization presented in the first chapter, let us discuss the properties of the most commonly used organic supports and how cell attachment is accomplished. The carriers that will be presented can be subdivided into carrier polysaccharides, proteins, and synthetic polymers.

3.3.1   Carrier Polysaccharides

A.   K-carrageenan[9,10]

Carrageenan is a polysaccharide consisting of β-D-galactose sulfate and 3,6 anhydro-α-D-galactose (isolated from seaweed). It becomes a gel under mild conditions and is thus one of the most suitable matrices for microbial immobilization. Carrageenan (0.05–20%) is dissolved in physiological saline, warmed to 37–50°C, and mixed with a microbial suspension also heated to the same temperature. The mixture is cooled and contacted with an aqueous solution containing gel inducing agents such as alkali metal ions (aluminum, magnesium, ferric, ferrous), organic amines (alkylenediamine, phenylenediamine, hydroxamate, hydraside), K^+, or NH_4^+ at the concentration of 0.1 mM.

The cell-polysaccharide mixture gels under these conditions. The final step involves contacting the matrix with gel-hardening agents such as diisocyanate, isothiocyanate, or carbodiimide, at a concentration from 0.01 to 1 gram per ml and at the temperature of 30°C. Afterwards, the gel is granulated into particles of a suitable size—the immobilized cell preparation is now ready. There are several advantages to this method: 1. It is simple 2. Various shapes of the microbial biocatalyst can be obtained (bead, cubic, membrane). 3. Immobilization does not affect the metabolic activity of the cells, which remain alive. Microbial cells immobilized into the carrageenan gel exhibit higher activity than the same cells attached to the acrylamide support. This suggests an important contribution of the support itself to the biocatalyst system. It was determined that the carrageenan matrix in its gel form exhibits a stabilizing effect on the immobilized cell preparation. It is possible that as a result of metal incorporation into the polysaccharide matrix, partial covalent bonds form between carboxyl and amino groups on the cell surface and the metal, and between the same metal and reactive groups on the support surface. Since some of these metal ions undergo oxidation-reduction, they may activate cell membrane proteins responsible for product excretion. As a result, cells immobilized into carrageenan gel are metabolitically very active and can carry out various single as well as multi step enzymatic reactions. Carrageenan catalysts are used for aspartic acid, alanine, isoleucine, fructose, uroconic acid, etc. production (Table 4).

B. Agar[11]

Agar is a polysaccharide consisting of β-1, 4- D-galactopyranoside and α-1,3,3,6- anhydro-α-galactopyranoside. It contains sulfur groups in the polymer and its properties are rather similar to properties of kapa-carrageenan. A *Saccharomyces pastorianus* cell suspension was entrapped into 2.5% (w/v) agar solution at 50°C, with a 1:4 volume ratio of cells to agar. When the mixture was quickly placed into a cold solution of toluene or tetrachloroethylene, pellets with spherical shapes were formed. In them, the yeast cells were distributed homogeneously. Oxygen and product diffusion limitations are the major disadvantages of this technique.

C. Celluloses

Celluloses represent inexpensive and stable supports with three hydroxyl groups on each anhydro glucose unit. Substituted celluloses are prepared by replacing the hydroxyl groups by various substituents. Replacement by diethyl-aminoethyl (DEAE) introduces amino groups, which provide polycationic, weakly basic anion exchangers. Replacement of hydroxyl by carboxy methyl

Table 4   Cells Immobilized into Carrageenan Gel

Microbial Strain	Enzyme	Substrate	Product
E. coli	Aspartase	Fumaric acid + ammonia	L-aspartic acid
A. oryzae	Aminoacylase	Hydrolysis of N-acyl DL amino acids	L-amino-acids
Pseudomonas docunhae	Aspartate β-decarboxylase	L-aspartic acid	L-alanine
Serratia marcescens	Multienzyme	Glucose	L-isoleucine
Serratia marcescens	Multienzyme	Glucose	L-histidine
B. subtilis	Multienzyme	Glucose	Arginine
Streptomyces griseus	Glucose isomerase	Glucose	Fructose
Streptomyces phaecochromogenes	Glucose isomerase	Glucose	Fructose
E. coli	Penicillin amidase	Penicillin	6-APA
Sarcina lutea	Urease	Urea	Ammonia + CO_2
Brevibacterium ammoniagenes	Fumarase	Fumarate	L-malic acid
Pseudomonas putidum	L-arginine deiminase	L-arginine	L-citrulline
Achromobacter liquidum	L-histidine ammonia lyase	L-histidine	Uroconic acid

(CM) substituents introduces carboxyl groups, which provide weakly acidic cation exchangers.

Regular cellulose has a fibrous shape and is thus a high surface area, porous material; however, it shows rather poor flow properties in column operation. A method of preparing highly porous carrier beads was patented by Tsao and Chen.[12] Any fibrous cellulose (mono, di, tri acetate, methyl, ethyl, carboxymethyl, diethylaminoethyl, mixed amines, phosphoric acid, etc.) can be converted to bead form by first dissolving it in an organic solvent (pentane, hexane, decane at ratios from 1:10 to 1:6 [w/v], respectively). The beads formed can be hardened by treatment with crosslinking reagents such as diisocyanate or aminopropyltrioxysilane. A review of microbial adsorption and coupling to various forms of celluloses follows.

Adsorption to Cellulose[13]

Spores of *Aspergillus oryzae* NRRL 1989, *A. wentii* NRRL 2001, *Penicillium requeforti* NRRL 3360, and *Micrococcus luteus* ATCC 9341 (Table 5) were immobilized by adsorption to fibrous carboxymethyl, phosphoric acid, diethylaminoethyl, and mixed amines. A measured amount of cellulose

Table 5   Exchange Groups of Celluloses Used as Microbial Supports (Immobilization via Adsorption)

Designation	Exchanger Group	pH Range Tested	Support Retention Capacity for			
			A. oryzae NRRL 1989	P. roquefortii NRRL 3360	A. wentii NRRL 2001	M. luteus ATCC 9341
CM Carboxymethyl	$\backslash$-O-CH$_2$-COO-H$^+$ (cation exchanger)	5	Good	Good	Good	none
		9	Good	none	Poor	—
P Phosphate	$\backslash$-O-P(O) O·H$^+$ O·H$^+$ (cation exchanger)	5	excel	excel	Good	—
		9	Good	Poor	Poor	—
DEAE Diethylaminoethyl	$\backslash$·O·C$_2$H$_4$·N$^+$H·(C$_2$H$_5$)$_2$OH$^-$ anion exchanger	5	Excellent	Poor	Poor	—
		9	Excellent	Good	Excellent	—
ECTEOLA Mixed amines	$\backslash$(undefined)OH$^-$ anion exchanger	5	Excellent	Excellent	Excellent	—
		9	Excellent	Excellent	Excellent	—

(approximately 8 grams for a 200 mm high column with a 15 mm diam.) was washed with several volumes of 0.05 M K_2HPO_4 − NaOH (pH = 5.8) buffer solution. Microbial cells were added to the cellulose support; the mixture was poured into a column and allowed to settle. Unadsorbed cells were washed out with the same buffer solution. For *A. oryzae, A. wentii,* and *P. requeforti* strains, the ecteola support (anion exchanger) was best, compared with other celluloses. For *Micrococcus luteus,* carboxymethylcellulose (the only support tested) was not the carrier of choice using adsorption as the method of attachment. This support practically retained no cells, and the composite formed exhibited no histidine ammonium lyase enzyme activity. *M. luteus* cells can be successfully immobilized by coupling to carbodiimide activated carboxymethylcellulose.

Coupling to Cellulose[14]

Micrococcus luteus (ATCC 9341) cells were covalently linked to carboxymethylcellulose activated with carbodiimide (EDAC). The immobilization reaction consisted of two steps: 1. activation of cellulose by carbodimide 2. cell attachment to activated cellulose. The mechanism of the reaction involved is presented in diagram 1: after carboxymethylcellulose was treated with EDAC, an activated cellulose intermediate (1) was formed. The activated intermediate

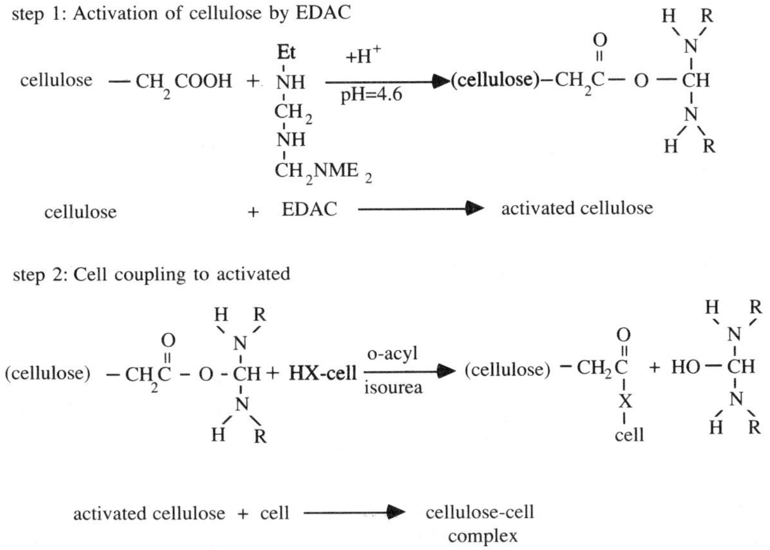

Diagram 1   Cell coupling to cellulose.

reacted with a nucleophilic group (HX) on the cell surface in the presence of o-acylisourea. As a result of this reaction, a covalent bond formed between the cellulose support and the cell. It is interesting to note that carboxymethylcellulose without activation by EDAC did not retain *M. luteus* cells at all, but after treatment, the biocatalyst formed did exhibit L-hystidine ammonium lyase enzyme activity. Several other supports such as Affi-Gel 201, Affi-Gel 202 were tested for *M. luteus* ATCC 9341 immobilization (Table 6). Apparently, the cellulose support Affi-Gel 202 was better than Affi-Gel 201, and both carriers were superior to carboxymethylcellulose. The supports differ in the length of spacer arm on the surface: The longer the spacer arm, the more bonds can be formed between the carrier and the cell, thus making the support retention higher and the biocatalyst better.

3.3.2   Carrier Proteins

A.   Collagen[15–17]

A method of microbial cell immobilization using collagen, a natural fibrous protein, was developed by researchers at Rutgers University (USA). Collagen is hydrophilic and swells in the presence of water. The mechanism of microbial immobilization to collagen involves formation of multiple ionic interactions, hydrogen bonds, and Van der Waals interactions between the cell and the glycine and amino acids of collagen. While individually weak, taken together these bonds form a stable support-cell biocatalyst.

For the purpose of immobilization, collagen (0.5–5.0%) was blended with acidified water at 4°C. The allowable cell to collagen ratio (dry weight basis) was between .2 and 3. After initial flocculation, the system cooled in a freezer for 5–10 min. The pH was monitored during ensuing base (acid) addition and mixing. When the pH exceeded 11.0 (below 3.0 for acid), the swollen collagen dispersion was blended two or three more times in 30 sec bursts at 20,000 RPM (using a magnetic stirrer). Between two consecutive blendings, the system cooled for 5–10 min to prevent any denaturation of collagen because of local heating. Cells were then added to the dispersion while the pH was still over 11.0

Table 6   Immobilization of *M. luteus* ATCC 9341 Cells to Various Supports

Support	Reactive Groups on the Support Surface
CM	$Cellulose–O–CH_2COOH$
Affi-Gel 201	$Cellulose–O–CH_2CONH\ NHCOCH_2OCH_2COOH$
Affi-Gel 202	$Cellulose–OCH_2CONH(CH_2)_3NH(CH_2)_3NHCO(CH_2)_2COOH$

(under 3.0 for acid) and the process of blending/cooling was repeated. A membrane formed by laying the dispersion out on a Mylar sheet and leaving it to dry overnight at room temperature with forced air circulation. The dried membrane was tanned by dipping it in 10% (w/v) glutaraldehyde solution (pH = 8.0) for 0.5–5 min. Finally, the support was washed and dried.

Crosslinking of the Cell-collagen Membrane

There are two methods of crosslinking (or tanning) the membrane for greater strength: Surface tanning and bulk tanning.

1. Surface tanning: With this method, the cell-collagen suspension is cast and dried. The resulting membrane is cut into smaller pieces, which are immersed into a glutaraldehyde batch of a specific concentration for a specific period of time. The membrane is then washed under running water for 1–2 hours to remove all extra glutaraldehyde.
2. Bulk tanning: Using this method, glutaraldehyde is added to the cell-collagen dispersion before it is cast into a membrane. The addition of glutaraldehyde causes formation of bridge-like bonds between the collagen and cells. As a result, hydrogen ions are liberated and a drop in pH of the dispersion occurs. To control this reaction, glutaraldehyde is added in small portions, with intermittent addition of a base for pH regulation. The dispersion is then cast on a pregreased Mylar sheet as quickly as possible. Since glutaraldehyde continues to crosslink the dispersion, its viscosity keeps increasing. The membrane formed only needs to be dried.

An Improved Technique

A 1.5% collagen dispersion was formed at pH = 11.0 and poured into dialysis beds (performed at 4°C) made of a material whose molecular weight cutoff was 3000. This insured passage of small molecules and ions, especially divalent ions like Ca^{+2} and Mg^{+2}, which cause collagen to coagulate between the pH of 3 and 11, but passage of collagen macromolecules with weights of around 300,000 was blocked. The base and other ions diffused out into the distilled water, raising its pH and lowering the pH of the collagen. The dialysis was discontinued when the basicity of the water outside the bags stopped rising. The collagen thus obtained was blended at 20,000 RPM for 30 sec and then cooled in a freezer. As previously, the process of cooling and blending was repeated several times, after which, prewashed cells were added to the collagen to give a 50% loading. The mixture was blended, cooled, cast, and dried at 35°C for 1.5–2 hours.

Collagen was successfully applied as a support for attachment of various cells used in primary as well as secondary metabolite production, nitrogen fixation, waste treatment, etc. (Table 7). Since the cell loading of the collagen carrier can be rather high, and because the microbe-carrier complex formed is stable, collagen is an excellent support for microbial immobilization.

3.3.3 Synthetic Polymers

Synthetic polymers are often used as supports for microbial attachment. Depending on the type of monomer unit in the polymer matrix, they can be subdivided into copolymers of styrene and divinylbenzene, polymers of acrylamide and N,N'-bismethylenacrylamide, polymers of phenol and formaldehyde or styrene and malic acid, polymers of toluene diisocyanate and polyether diol (urethane polymers) (Table 1). The molecular structures of corsslinked styrene-divinylbenzene copolymers and acrylamide polymers are presented in Diagrams 2 and 3.

Table 7   Cells Immobilized into Collagen Support

Microbial Cells	Substrate	Products	Reaction Catalyzed
Serratia mercescens	Glucose	2-keto-gluconic acid	Multienzyme
Acetobacter sp.	Ethanol	Acetic acid	Multienzyme, cofactor
Corynebacterium lilium	Glucose	Glutamic acid	Primary metabolite
Aspergillus niger	Sucrose	Citric acid	Primary metabolite
Anacystis nidulans	Water	Oxygen	Immobilized algal cells
Anacystis nidulans	Nitrate	Ammonia	Biological nitrogen fixation
Streptomyces griseus	Glucose	Candicidin	Secondary metabolite, antibiotic synthesis
Pseudomonas aeroginosa	—	—	Waste treatment (bioadsorption of plutonium)
Klebsiella pneumoniae	Nitrogen	Ammonia	Microbial fixation of atmospheric nitrogen
E. coli ATCC 11303	Fumaric acid	Aspartic acid	Single enzyme
Corynebacterium simplex	Hydrocortisone	Prednisolone	Steroid transformation, single enzyme with co-factor requirement
Chloroplast	Water	Oxygen	Immobilized organelle biophotolysis of water
Mammalian erythrocyte	—	—	Model for studying in vivo enzyme reaction

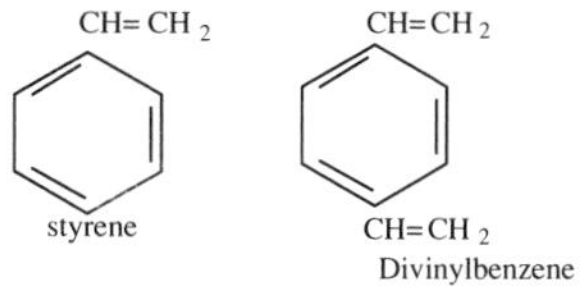

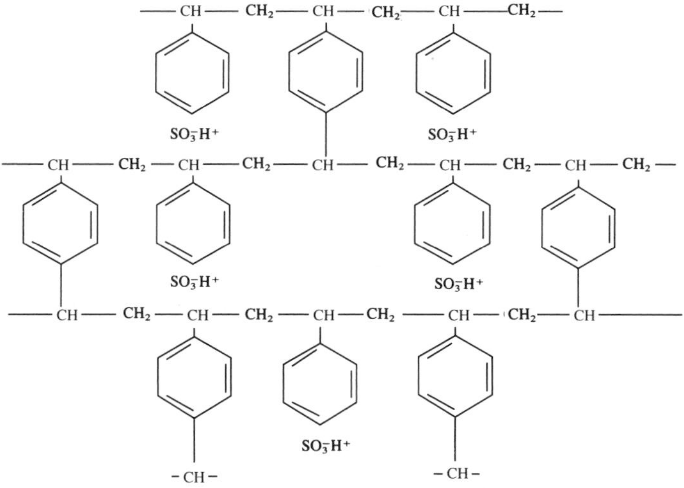

Diagram 2   Structure of styrene-divinylbenzene copolymer.

Divinylbenzene links together the straight chain of the styrene copolymer while N,N'-bismethylenacrylamide does the same to the straight chain of the acrylamide polymer. These compounds are also called "crosslinkage' reagents. The term "crosslinkage' refers to the percentage of divinylbenzene or N,N'-bismethylenacrylamide in the polymer matrix. Crosslinkage is a parameter which refers to the mechanical strength, porosity, swelling and shrinking of the support. Often a polymer matrix contains a large number of ionizable groups bonded to it, which are located throughout the entire ion exchange resin structure. Depending upon the types of diffusible groups in the support matrix, the ion exchange resins are divided into cation and anion exchangers. The methods used for microbial attachment to the polymer supports are adsorption and entrapment.

A.   Adsorption to Ion Exchange Resins[18]

Microbial cells exhibit a dipolar character, behaving as cations or anions depending on the pH of the solution. Charged microbial cells can interact

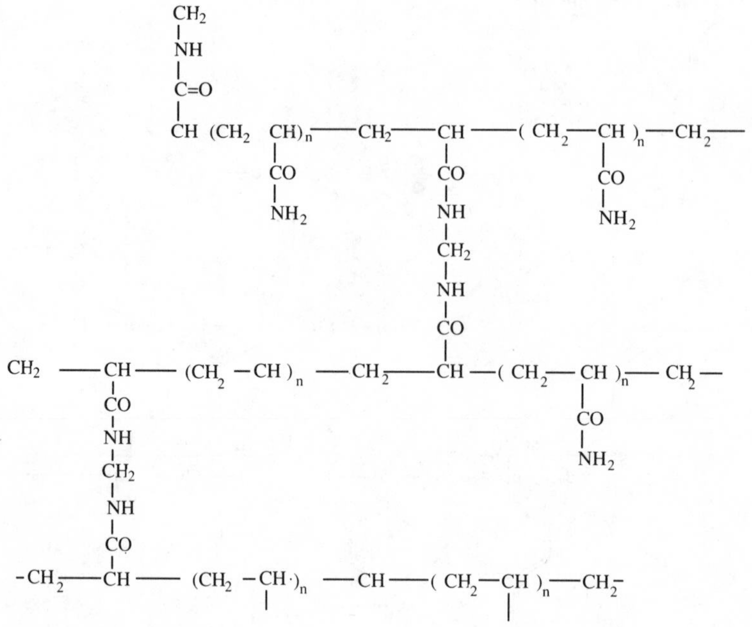

Diagram 3 Structure of acrylamide polymer.

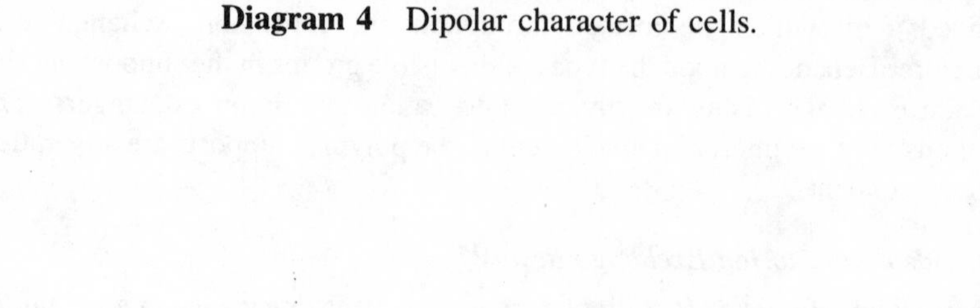

Diagram 4 Dipolar character of cells.

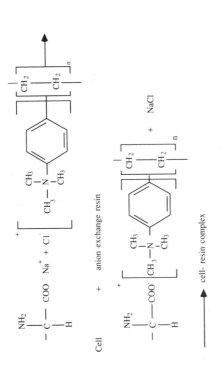

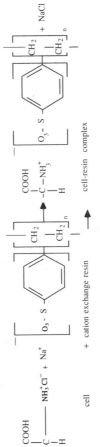

Diagram 5 Mechanism of microbial attachment to ion-exchange resins.

electrostatically with cation or anion exchange resins, forming stable microbe-carrier complexes, as shown in Diagram 5.[19]

A microbial cell suspension was circulated through a column with resin, and the latter washed by decantation to remove nonattached cells. It was found that Dowex resin adsorbed *Pseudomonas aeroginosa*, *E. coli*, and *Saccharomyces carlsbergenis* cells tightly. Within the first hour of incubation there were no free cells on the resin surface. Table 8 lists other microorganisms adsorbed onto cation and anion exchange resins. The factors affecting microbial attachment to these materials include carrier properties, presence of specific ions, initial cell concentration, method of cell preparation prior to adsorption, as well as pH of the solution. It was determined that microbes immobilized by adsorption onto resins remained alive; moreover, several changes in cell metabolism were noted:

Nitrobacter agilis cells were attached to three types of Dowex 50 W-X sulfonated resins: polystyrene-divinylbenzene (PSD) with an ion exchange capacity (IEC) of 6.3 meg/gr, PSD with IEC of 0.007 meg/gr, and PSD with IEC of 0.074 meg/gr. The cells showed a shift in pH optimum (0.5 units higher) and a high rate of nitrification (100%) after immobilization onto all three supports.[20]

Azotobacter agile ATCC 9046 microbes, adsorbed onto the anion exchanger Dowex 1[21] resin (chloride form) for succinate oxidation, displayed a shift in pH optimum (one unit higher), twenty five percent higher glucose oxidase enzyme activity, and disappearance of lag time compared with free cells.

Pseudomonas aeroginosa cells were also immobilized onto anion exchanger Dowex 1, chloride form.[22] After attachment, the catalyst was placed into a continuously pumped growth medium [1% glucose, 0.5% $(NH_4)_2SO_4$, salts, pH = 7.0]. Two peaks in cell concentration of the medium were observed; the first, (a gradual peak) appeared after two hours of incubation. The authors attributed it to desorption of some cells from the carrier surface. The second, (rapid) peak in cell concentration occurred after 24 hours of growth. It was mainly attributed to detachment of adsorbed cells, which were able to grow on the resin surface. The rate of cell increase during the second peak reflected the growth rate of adsorbed cells on the support surface. The study also showed that adsorbed cells had a higher respiration activity. These data suggest that the carrier had a direct influence on the physiological character of the cells.

B. Entrapment into Acrylamide

Application of acrylamide for microbial immobilization was first described in 1974.[23] *Brevibacterium ammoniagenes*, *Corynebacterium equi*, *Escherichia coli*, *Microbacterium flavum*, *Proteus vulgaris*, *Bacillus* sp., etc. were suspended in small amounts of saline solution, acrylamide monomer, bismethylenacrylamide, initiator, and accelerator of polymerization. The reaction, a

Table 8 Cells Immobilized into Ion Exchange Resins

Microorganism	Type of Resin	Retention Capacity	References
Azotobacter vinelandii	Anion exchanger	0	B. Rotman, *Bacterial Review*, 1960, 24, 2, 251
Escherichia coli	Anion exchanger	4	B. Rotman, *Bacterial Review*, 1960, 24, 2, 251
Escherichia coli B	Anion exchanger	100	B. Rotman, *Bacterial Review*, 1960, 24, 2, 251
	Cation exchanger	19	B. Rotman, *Bacterial Review*, 1960, 24, 2, 251
Escherichia coli K-12	Anion exchanger	0	B. Rotman, *Bacterial Review*, 1960, 24, 2, 251
	Cation exchanger	0	B. Rotman, *Bacterial Review*, 1960, 24, 2, 251
Salmonella schottmulleri	Anion exchanger	23	B. Rotman, *Bacterial Review*, 1960, 24, 2, 251
Salmonella typhosa	Anion exchanger	2	B. Rotman, *Bacterial Review*, 1960, 24, 2, 251
Shigella dysenteria	Anion exchanger	0	B. Rotman, *Bacterial Review*, 1960, 24, 2, 251
Staphylococcus aureus	Anion exchanger	17	B. Rotman, *Bacterial Review*, 1960, 24, 2, 251
Escherichia coli Yamagashi strain	Anion exchanger (Dowex 1)	Not available	R. Hattori, et al., *Bacterial Review*, 1972, 18, 271, 285
Pseudomonas aeroginosa	Anion exchanger	Not available	R. Hattori, et al., *Bacterial Review*, 1972, 18, 271, 285
Nitrobacter agilis	Anion exchanger (Dowex 50)	Not available	R. Hattori, et al., *Bacterial Review*, 1972, 18, 271, 285
Serratia marcescens	Anion exchanger		Zwejagintzev, *Microbiologya*, 1971, XL, 1, 139
Staphylococcus aureus	Anion exchanger		Zwejagintzev, *Microbiologya*, 1971, XL, 1, 139
Bacillus cereus	Dowex 1		Zwejagintzev, *Microbiologya*, 1971, XL, 1, 139
Bacillus mycoides	Dowex 1		Zwejagintzev, *Microbiologya*, 1971, XL, 1, 139
Bacillus megaterium	Dowex 1		Zwejagintzev, *Microbiologya*, 1971, XL, 1, 139
Bacillus subtilis	Dowex 1		Zwejagintzev, *Microbiologya*, 1971, XL, 1, 139

free radical polymerization, took 15 minutes to complete at 25°C. Since all reagents are toxic to microbes, cell activity was severely damaged. To reduce this effect, the influence of the total concentration of all reagents and their relative concentrations should be carefully investigated for each particular microorganism.[24] The acrylamide: bismethylenacrylamide ratio determines mean pore size of the gel in which microbial cells are entrapped and the fragility of the matrix formed. For example, when the concentration of bisacrylamide monomer is higher than 20%, the gel becomes fragile.

The free radical polymerization reaction liberates heat. To reduce cell thermoinactivation, the column temperature should be carefully controlled. There are different points of view on the microbial state after immobilization. A widely held opinion is that immobilization into acrylamide results in cell destruction, but not inactivation of enzymes, which remain capable of carrying out various reactions for a period of time (Table 9). Due to polyacrylamide toxicity, many researchers have now switched to other attachment techniques, especially entrapment into the carrageenan gel. A comparison of enzymatic activity and operational stability of microbial cells immobilized onto acrylamide and carrageenan gels showed that the carrageenan support exhibits higher enzyme activity and a longer operational half-life.[10] This result emphasizes the importance of support contribution to the biocatalyst system.

C.   *Entrapment into Photo-crosslinkable Prepolymers and Polyurethane Matrices*

These methods of cell immobilization were developed by scientists at Kyoto University in Japan.

1.   *Entrapment into Photo-crosslinkable Polymers.*   The polymers are formed during brief illumination of photocrosslinkable prepolymers. Various prepolymers are available as starting materials for gel matrices, including polybutadiene (PB-200K), maleic polybutadiene (PBM-2000), and polypropylene glycol (ENT-2000).

Microbial cells were suspended in a water-acetone mixture containing a prepolymer. The mixture was spread on a sheet of transparent polyester and irradiated from a UV light source for 3 min. Polymerization occurred, forming a matrix containing entrapped cells. The gel was then cut into small pieces (5 × 5 mm) and used as a catalyst for various transformations.

2.   *Entrapment into Polyurethane Matrices.*   Polyurethane matrices are synthesized from toluene diisocyanate and polyester diol. They differ in molecular weight, NCO content of the prepolymer, and ethylene oxide content in the diol, Table 10. The method of cell immobilization is entrapment. However, the mechanism of cell attachment is entrapment and coupling since it is believed that covalent bonds form between amino groups on the cell surface

Table 9 Cells Immobilized into Acrylamide Gel

Microbial Cell	Enzyme	Substrate	Product	Operational Half-Life	
				Days	Temperature
Pseudomonas putida	L-arginine deiminase	L-arginine	L-citrulline	140	37°C
Achromobacter liquidum	L-histidine ammonia lyase	L-histidine	Uroconic acid	180	37°C
Escherichia coli	Penicillin amidase	Penicillin	6-APA	42	30°C
Achromobacter aceris	NAD-kinase	NAD, ATP	NADP, ADP	20	37°C
Brevibacterium ammoniagenes	Polyphosphate NAD kinase	NAD, $(Pi)_n$	NADP, $(Pi)_{n-1}$	15	36.5°C
Achromobacter butyri	Polyphosphate glucose kinase	Glucose $(Pi)_n$	Glucose-6-phosphate	20	37°C
Saccharomyces cerevisiae	Glutathione synthetase	L-glutamic acid, L-cysteine, glycine	Glutathione	20	30°C
Aspergillus niger		Glucose	Gluconic acid		
Proteus rettgeri	Penicillin amidase	Penicillin	6-APA		

Table 10 Photo-crosslinkable* and Polyurethane† Matrices

Prepolymer Type	Molecular Weight of Prepolymerdiol	NCO Content (%) of Prepolymer	Ethylene Oxide Content (%) in Polyether diol	Reaction Catalyzed (Enzyme)	References
PB-200K*				Steroid dehydrogenation (enzyme + cofactor)	1
PBM-2000*				Steroid dehydrogenation (enzyme + cofactor)	1
ENT-2000*				Steroid dehydrogenation (enzyme + cofactor)	1
PU-1†	1486	4.0	100	Steroid dehydrogenation (enzyme + cofactor)	2
PU-2†	2529	3.1	57	Steroid dehydrogenation (enzyme + cofactor)	2
PU-3†	2529	4.2	57	Steroid dehydrogenation (enzyme + cofactor)	2
PU-4†	2529	5.6	57	Steroid dehydrogenation (enzyme + cofactor)	2
PU-5†	2627	2.7	91	Steroid dehydrogenation (enzyme + cofactor)	2
PU-6†	2627	4.0	91	Steroid dehydrogenation (enzyme + cofactor)	3
PU-7†	2627	5.6	91	Steroid dehydrogenation (enzyme + cofactor)	3
PU-8†	2616	2.7	100	Steroid dehydrogenation (enzyme + cofactor)	3
PU-9†	2616	4.0	100	Steroid dehydrogenation (enzyme + cofactor)	3
PU-10†	2616	5.6	100	Steroid dehydrogenation (enzyme + cofactor)	3
PU-11†	4285	4.0	73	Steroid dehydrogenation (enzyme + cofactor)	3
Hydrophilic polyurethane					
Hypol® 2000				6-APM (penicillin acylase)	4
Hypol® 3000				6-APM (penicillin acylase)	4
Desmodur T80 FRL 2235				Single enzyme penicillin acylase for 6-APA	4

Hypol†	Isocyanate Content (mg/gr)	Activity (unit/gr)			
A	2.16	48,966		Single enzyme (aspartase)	5
B	1.63	61,035		Single enzyme (aspartase)	5
C	2.48	46,035		Single enzyme (aspartase)	5
D	1.79	68,017		Single enzyme (aspartase)	5

[1] T. Omata, et al., Applied Microbiology and Biotechnology, 1979, 6, 207; [2] K. Sonomoto, et al., Agricult. Biol. Chem., 1980, 44, 5, 1119; [3] T. Omata, et al., Journal of Fermentation Technology, 1980, 58, 4, 339; [4] J. Klein and M. Kluge, Biotechnology Letters, 3, 2, 65; [5] M. Fusee, et al., Applied Environmental Microbiology, 1981, 42, 673.

and isocyanate groups of the matrix. These supports are sold in liquid and solid forms.

A solid urethane matrix was melted at 60°C and cooled to 37°C. Cells in a buffer were mixed with the melted matrix, layered on a glass plate, and allowed to stand for one hour at 4°C. The resin formed was cut into pieces and used as a catalyst.

The liquid form of the matrix was quickly mixed with cells and water. Polycondensation started immediately and the reaction mixture swelled to five times its volume during the first five minutes. There was no pH shift and only a moderate temperature increase during the reaction, which concluded in 20 min. By varying the polyisocyanate: H_2O ratio, supports of different porosities and densities were formed; higher water content lead to matrices with smaller pores and increased density. Depending on the type of polyisocyanate, (Hypol® 3000, Hypol® 2000, Desmodur T80 FRL 2235), and the water: prepolymer ratio, (from 0.3:1.0 to 7.0:1.0 w/w), differently structured catalysts formed: foam, gel, and spherical foam particles. The gel synthesized from Desmodur type polyisocyanate had a porous structure and good substrate/product flow, which allowed cell immobilization inside the pores. However, increasing the H_2O/ prepolymer ratio decreased the pore size of the gel.

The disadvantage of this type of matrix is isocyanate toxicity to the cells and the adverse effect high isocyanate content may have on catalyst performance. But this support matrix also has numerous advantages:

1. The method of immobilization is simple;
2. Matrices with various forms can be made;
3. By combining hydrophobic and hydrophilic polyurethane matrices, a support most suitable for a particular transformation can be tailored;
4. The supports can withstand high pressures (100–500 psi).

The matrices described are now widely used for 6-APA and aspartic acid production, as well as for various steroid transformations.

Literature

1. Chibata, I., Tosa, T., *TIBS*, April, 88 (1980)
2. Vijayalakshmi, *et al.*, *Ann. NY Acad. Sci.*, **249** (1979)
3. Vieth, W. R., Venkatasubramanian, K., *ACS Symposium*, Series 106, 1 (1979)
4. Durand G., Navarro, J., *Process Biochemistry*, September, 14 (1978)
5. Marcipar, *et al.*, *Biotech Lett.* **1**, 2, 65 (1980)
6. Messing, R. A., Oppermann, R. A., and Kolot, F. B., *ACS Symposium*, Series 106, 10 (1979)
7. Kolot, F. B., *Developments in Industrial Microbiology*, **21**, 295 (1980)
8. Kolot, F. B., *Process Biochemistry*, Oct./Nov., 2 (1980)
9. Chibata, I., Tosa, T., Tokata, I., US Patent 4,138,261 (1979)
10. Chibata, I., *ACS Symposium*, Series 106, 13 (1979)
11. Toda, K., *Biotech. Bioeng.*, **17**, 487 (1975)

12. Tsao, G. T., Chen, L. F., US Patent 4,063,017 (1977)
13. Johnson, D. E., Ciegler, A., *Archiv. Biochem. Biophys.*, **130**, 384 (1969)
14. Jack, T. R., *Biotech. Bioeng.*, **XIX**, 631 (1977)
15. Vieth, W. R., *et al.*, US Patent 3,972,776 (1976)
16. Vieth, W. R., *et al.*, *Biotech. Bioeng.*, **XV**, 565 (1973)
17. Venkatasubramanian, K., *et al.*, *J. Ferm. Techn.*, **52**, 4, 268 (1974)
18. Rotman, B., *Bacteriological Review*, **24**, 2, 251 (1960)
19. Daniels, S. L., *Developments in Industrial Microbiology*, **13**, 21, 1972.
20. McLaren, A. D., Skujins, J. J., *Can. Journal of Microbiology*, **9**, 729 (1963)
21. Hattori, T., Furusaka, C., *J. Biochemistry*, **50**, 4, 312 (1961)
22. Hattori, T., *et al.*, *J. of General and Applied Microbiology*, **18**, 272 (1972)
23. Chibata, I., *et al.*, *J. of General and Applied Microbiology*, **27**, 878 (1974)
24. Suzuki, S., Karube, I., *ACS Symposium*, Series 106, 59 (1979)
25. Navarro, J. M., Durand, G., *European J. Applied Microb.*, **4**, 243, 1977
26. Navarro, J. M., *SMT Coloque*, Toulouse, France, 211 (1978)
27. Mole, M., *et al.*, US patent 4,009,286 (1977)
28. Weetall, H., *et al.*, *Biotech. Bioeng.*, **XVI**, 689 (1974)
29. Barker, S. A., *et al.*, US Patent 3,912,593 (1975)
30. Kennedy, J. F., *et al.*, *Nature*, **261**, 242 (1976)
31. Helmstetter, C. E., *J. Mol. Biology*, **24**. 417 (1967)
32. Helmstetter, C. E., Cooper, F., *J. Mol. Biology*, **30**, 507 (1968)
33. Thonart, P., *et al.*, *Enzyme and Microbe Technology*, **4**, 191 (1982)
34. Van Haecht, J., *Biotech. Bioeng.*, **27**, 3, 217 (1985)
35. Adlercreutz, P., *Appl. Micro & Biotech.*, **22**, 1, 1 (1985)
36. Horitsu, H. *et al.*, *Appl. Micro & Biotech.*, **22**, 1, 8 (1985)
37. Schnaar, R. *et al.*, *Anal. Biochem.*, **151**, 2, 268 (1985)
38. Zueva, N., *Prikl. Biokhim.*, **21**, 3, 333 (1985)
39. Adlercreutz, P. *et al.*, *Appl. Micro & Biotech.*, **20**, 5, 296 (1984)
40. Mauz, O. *et al.*, *Ann. NY Acad. Sci.*, **434**, 251 (1984)
41. Krysteva, M. *J. Appl. Biochem.*, **6**, 5–6, 367 (1984)
42. Lobarzewski, J., *Biochem. Biophys. Res. Commun.*, **121**, 1, 220 (1984)
43. Tysiachnaia, I., *Prikl. Biokhim. Mikrobiol.*, **20**, 1, 79 (1984)
44. Kuu, W. *et al.*, *Biotech. Bioeng.*, **25**, 8, 1995 (1983)

4

Inorganic Supports for Microbial Attachment

The topic of Chapter 4 continues to be support criteria, strategies for carrier selection, and mechanisms of microbe-carrier interactions, but the focus shifts from organic to inorganic supports. Since adsorption and coupling are the major methods of immobilization to this type of carrier, special attention will be given to cell/support surface charges and how they are affected by pH, as well as to reactive groups on carrier surfaces.

Introduction

Although inorganic support materials have fewer reactive groups on their surfaces compared with organic supports, historically, they have been more widely used for microbial attachment (Table 1). Inorganic carriers can be subdivided into ungrafted and specially grafted supports (inorganic materials with specific organic groups attached by various coupling agents to their surfaces). The techniques used for microbial immobilization to inorganic carriers are adsorption and coupling. Three major mechanisms are responsible for cell attachment: 1. Electrostatic interactions between charged cell and charged carrier, 2. Partial covalent bond formation via replacement of hydroxyl groups on inorganic support surface with amino or carboxyl groups on cell surface, 3. Covalent bond formation between ligands on the cell surface and organic groups specially grafted to the inorganic carrier. For this reaction to occur, the inorganic surface should first be treated with a so called coupling agent.

Table 1 Cells Immobilized to Inorganic Supports

Microbial Strain	Support	Substrate	Products	Reaction Catalyzed	References
Saccharomyces carlsbergensis	Porous silica	Glucose	Ethanol	Multi-enzyme	25
Saccharomyces cerevisiae	Brick	Glucose	Ethanol	Multi-enzyme	26
Saccharomyces carlsbergensis	Brick	Wort	Beer	Multienzyme	27
Candida tropicalis	Biodamine ceramic	Glucose	Respiration rate		5
Saccharomyces cerevisiae	Biodamine ceramic				5
Rhodotorula sp.	Biodamine ceramic				5
Trichosporon sp.	Biodamine ceramic				6
Escherichia coli	Fritted glass, cordierite and zirconia ceramics	Glucose	Biomass accumulation	Multienzyme	6
Serratia marcescens *B. subtilis* *A. niger* *S. cerevisiae*	Fritted glass, cordierite and zirconia ceramics	Glucose	Biomass accumulation	Multienzyme	6
Penicillium chrysogenum	Fritted glass, cordierite and zirconia ceramics	Glucose	Biomass accumulation	Multienzyme	6
Streptomyces olivochromogenes	Fritted glass, cordierite and zirconia ceramics	Glucose	Biomass accumulation	Multienzyme	6
E. coli	Metal hydroxides				28
S. cerevisiae	Metal hydroxides				28
Serratia marcescens	Metal hydroxides				28
Lactobacillus	Metal hydroxides	Glucose	Acetic acid	Multienzyme	28
Acetobacter	Metal hydroxides	Glucose	Acetic Acid	Multienzyme	28

A great variety of inorganic supports such as sand, brick, glass, ceramics, mineral silicates, metal oxides, magnetic particles, etc., have been utilized as supports for microbial attachment. In spite of the fact that adsorption to an inorganic support is generally the simplest and quickest procedure, the strength of microbial attachment greatly depends on cell wall composition as well as on carrier surface properties, and is affected by the pH, ionic strength of the solution, cell age, surface charges, surface area of the support, and carrier composition. Each of these factors will be discussed in detail.

4.0   Factors Influencing Microbial Adsorption

4.0.1   Effect of pH

The effect of pH on microbial attachment was studied by several authors. Using *Saccharomyces carlsbergensis* yeasts, Navarro[25] investigated cell immobilization to three supports (kieselguhr, bentonite-H^+, and amine bentonite). This worker found that kieselguhr and bentonite-H^+ adsorbed significantly at pH = 3, while amino bentonite did the same at pH = 5. The data suggest that simply changing the pH induces an important variation in carrier retention properties as a result of modification of the support's microenvironment.

Marcipar *et al.*[5] studied the adsorption of four microorganisms (*Trichosporon sp.*, *Rhodotorula sp.*, *Saccharomyces cerevisiae*, and *Candida tropicalis*) to the inorganic Biodamine support as a function of two different pH values, 4 and 6. Cells of the different microbial strains were adsorbed to the support from a standardized suspension with a carefully defined microbial concentration. The results demonstrate that although the rate of attachment is rather specific for each strain tested, the percentage of cell fixation was higher at a pH of 4.0 for all microbes. Almost a 40% decline in the amount of cell adsorption was noticed at a pH of 6.0.

4.0.2   Effect of Ionic Strength[2]

Ionic strength plays an important role in microbial adsorption. At a constant pH value, increasing ionic strength of the medium resulted in a higher percentage of cell fixation. At a pH of 4.0 and without addition of NaCl, 8.86% of the cells were immobilized. With addition of NaCl (forming a 2M solution), the number of cells immobilized increased to 16.03%. Greater microbial binding due to ionic strength of the medium was also observed at a pH of 6.0. The data thus confirm that microbial attachment by adsorption requires that special attention be paid to the pH and ionic strength of the solution. Even small modifications could result in microenvironmental changes, which affect charges on cell and support surfaces, ion-ion interactions, or silonol-cell partial covalent bond formation, and could cause cell desorption.

4.0.3   Effect of Zeta Potential (Thonart, 1982)

Zeta potential, the charge on a cell surface, plays a critical role in microbe attachment. Knowledge of the exact charge on the cell membrane can help in choosing a proper support.

The zeta potential of *S. cerevisiae* was studied as a function of culture age, pH, type of medium, concentrations of various ions [$CaCl_2$, NaCl, $Al_2(SO_4)_3$, etc.], and type of support.

It was shown that the potential on the microbial surface was independent of cell age, but that the overall charge of haploid cells ($-10mV$) was on the average two times higher compared with diploid cells ($-5mV$).

Most microbes have a dipolar character which can function as a cation or anion depending on the pH of the solution:

Depending on the pH of the solution, a cell can therefore change its functional properties.

Raising the pH from 3.0 to 7.5 resulted in an increase of cell charge from $-10mV$ to $-30mV$. The cell charge diminished with decreasing pH.

Depending on the ions present, their valency and the medium, the actual charge on *S. cerevisiae* surface varied in different directions. While various concentrations of sodium ions did not affect the cells' zeta potential, an increase in Ca-ion concentration from 0.01% to 1.0% resulted in an eight fold decrease of charge on the cell surface in a minimal medium. Placed in distilled water, the negative charge on yeast membrane increased five fold ($-7.8mV$ to $-39.3mV$). However, increasing the concentration of Ca^{++} resulted in the cell surface decharging.

Experiments with Tween 80 and $Al_2(SO_4)_3$ gave similar results. In the presence of high concentrations of carboxymethylcellulose, the negative zeta potential of *S. cerevisiae* quadrupled from $-7.8mV$ to $-29.8mV$. The most interesting effect on cell charge was that of cationic starch, increasing the concentration of which in a minimal medium resulted in reduced electric potential on the *S. cerevisiae* yeast surface (from $-7.8mV$ to $-2.0mV$). As noted, the zeta potential of yeast in distilled water increased from $-7.8mV$ to

−39.3mV, but adding a very small amount of cationic starch (1 mg/ml, 0.1%) resulted in the cell surface recharging to +26.5mV. Under similar conditions, the zeta potential on sawdust was measured as −12mV. This result shows that under specific experimental conditions, the originally negatively charged *S. cerevisiae* cells can assume a positive charge, resulting in strong cell adsorption to the carrier and therefore a stable microbe-carrier complex.

The above data suggest that various ions, pH of the medium, and medium composition can influence the charges on cell as well as on support surfaces. Knowledge of the zeta potential under specific experimental conditions can lead to improvement of support retention capacity and may result in better catalyst performance in the reactor.

4.0.4 Effect of Support Surface Charge

It is obvious that negatively charged bacterial or yeast cells would readily attach to the surfaces of positively charged supports such as DEAS-Sephadex A-50, DEAE-Sephadex A-25, Amberlite IR-45, and Biorad AG-21K via electrostatic interactions. But it appears that negatively charged cells can also be successfully attached to negatively charged inorganic carriers like glass and ceramics. In order to understand the mechanisms involved in microbe-carrier interactions, it is necessary to know the exact charge on the support surface. This can be found by studying the electophoretic mobility of various supports using the electrophoretic mass-transport analyzer. The principle behind this method is based on migration of charged carrier particles into or out of the particle chamber, depending on the carrier charge and on polarity of the chamber electrode. Change in support weight inside the chamber can be determined gravimetrically. From it, the electrophoretic mobility and electric potential on the support surface can be calculated. All five inorganic carriers under investigation were negatively charged, but the magnitudes of these charges were quite different. The smallest potential was observed on the surface of fritted glass. Inorganic carriers designed as cordierite had potentials almost equal to the charge of the natural support pumice, whereas zirconia coated ceramic had almost six times more negative charge on its surface compared with other carriers. This support also exhibited highest biomass accumulation of *Penicillium chrysogenum*, and the biocatalyst formed remained stable during long term continuous column operation. These data are in correlation with high lactase enzyme loading on the zirconia coated porous glass support, as was earlier observed.[28]

The high biomass accumulation of negatively charged cells on a negatively charged support suggests that charge-charge interactions cannot be the only mechanism involved in microbial attachment to an inorganic carrier.

4.1   Direct Bridge Formation between Inorganic Carrier and Cell

One possible reason for attachment of negatively charged cells to negatively charged supports may be partial covalency formed during immobilization. The creation of direct linkages, via chelation, between hydroxides of the transition metals (titanium, zirconium, iron, etc.) and microbial cells, was first described by Barker et al.[29] and Kennedy et al.[30]. Metalic hydroxides in aqueous solution usually undergo hydrolysis and polymerization, reducing the support surface charge. The degree of this polymerization is inversely proportional to the medium pH. The authors postulated that the mechanism of microbial attachment to metalic hydroxides involves replacement of hydroxyl groups on the surface of the supports by carboxyl, hydroxyl, or amino groups on the cell surface, which suggests a bonding through the mechanism of charge transfer between the transition metal and the microbial cell wall. The proposed structure of the zirconium hydroxide-amino group complex is illustrated in Diagram 1.

Escherichia coli, Saccharomyces cerevisiae, Serratia marcescens, Lactobacillus, and *Acetobacter* strains were immobilized on metalic hydroxides in the following manner: Microbial cells suspended in 0.9% w/v saline (A_{600} =

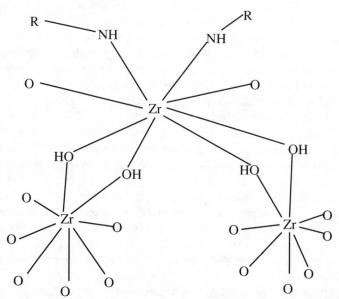

Diagram 1   Complexes of zirconium hydroxide with amino groups on the cell surface.

0.216) were mixed with a freshly prepared sample of metal hydroxide and agitated for five minutes at room temperature. The mixture was allowed to stand while the suspension settled out, leaving clear supernatant ($A_{600} = 0.022$). The immobilized cell preparation was then harvested by centrifugation. The coupled cells were alive, which was confirmed by measurement of cell respiration, and the authors proved that the microorganisms were bound to the surface of the hydroxide, not just trapped into the matrix. The authors recommended using zirconium hydroxide when applying microbial cells in continuous processes with pH ranges greater than 5.0. Titanium was recommended for processes with pH ranges from 2.0 to 5.0.

4.1.1 Effect of Support Composition on Microbial Coupling

Formation of partial covalent bonds, one of the mechanisms responsible for microbial attachment, suggests the importance of support composition. Inorganic carriers usually consist of a variety of oxides, such as aluminum, silicon, magnesium, titanium, etc. The chemical composition of one such ceramic support[5] commercialized in France and used for yeast immobilization is as follows: SiO_2—57.7%; Al_2O_3—38.1; Fe_2O_3—1.5%; TiO_2—1.4%; K_2O—0.5; Na_2O—0.1; CaO—0.4%; MgO—0.1%. (Ceramics as well as glass are considered hard liquids, thus, when placed into aqueous solution, ion exchange usually occurs on their surfaces.) Since microbial immobilization is performed in a buffer solution, metal hydroxides are formed on the surfaces of the inorganic supports instead of metal oxides. The hydroxyl groups of metal hydroxides can be replaced by suitable amino or carboxyl groups on the cell surface, as previously described for the zirconium complex. As a result, partial covalent bonds form between cells and the inorganic carrier.

Penicillium chrysogenum and *Streptomyces olivochromogenes* fungi were grown on a variety of inorganic supports, including silica glass, cordierite, and zirconia ceramic. Mycelial bioaccumulation on the supports was recorded after 24 and 48 hours of growth. It appeared that *P. chrysogenum* preferred cordierite over fritted glass. By contrast, *S. olivochromogenes* mycelium seemed to prefer fritted glass to cordierite after twenty four hours of incubation, and both fungi under investigation grew equally well on zirconia coated ceramic. These data understate the great importance of support composition for mycelial bioaccumulation.

A logical question would be whether it is possible to increase retention capacity of any carrier as well as long term biocatalyst activity by incorporating specific metals into the carrier matrices. Vijayalakshmi[2] used cross-linked pectate as the polysaccharide backbone, and coupled half of the support via iminodiacetate, while the other half was left without coupling. The coupled and

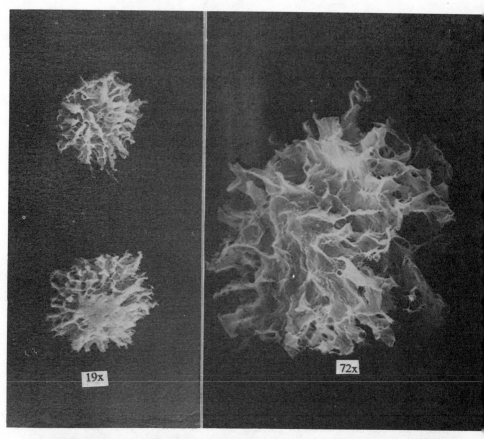

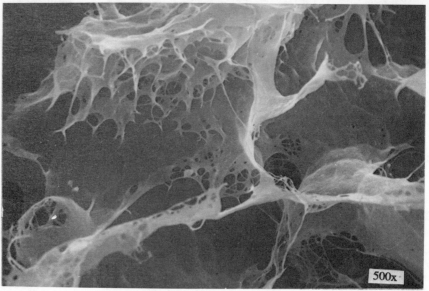

Scanning electron micrographs of *Streptomyces olivochromogenes* cells immobilized by partial covalent bond formation to inorganic carrier.

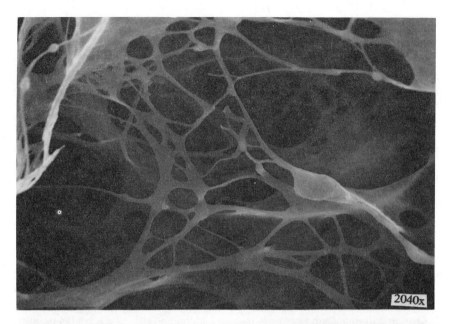

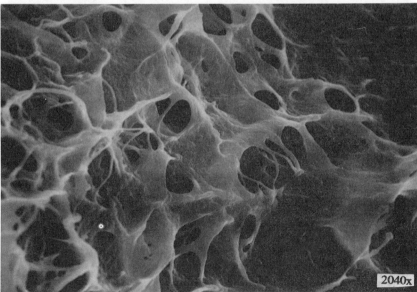

Scanning electron micrographs of *Streptomyces olivochromogenes* cells immobilized by partial covalent bond formation to inorganic carrier.

noncoupled pectate particles were suspended in a 0.5M $FeCl_3$ solution (pH = 5.0) for 24 hours. The gel was then washed free of excess metal and the Fe^{+3} content in the coupled support was determined (500μM/gr dry gel). In non-coupled pectate, the transition metal ions were attached only through the native carboxylic groups of the matrix and their concentration was lower. It was established that the backbone polysaccharide, with or without the crosslinking agent (iminodiacetate), but without Fe^{+3} did not bind the yeast cells at all. However, a carrier with Fe^{+3} retained almost 50% of the cells. Thus, after Fe^{+3} incorporation, the retention capacity of the pectate support became equal to that of the collagen support, which is a very good carrier. Incorporation of Ti or Zr metals into the same matrix resulted in supports with very low retentions. These data clearly show that adding specific ions to the carrier can increase the retention capacity of the support.

Chibata et al.[7] also used the organic polysaccharide carrageenan as a carrier for microbial attachment. After cell entrapment into the support, gelation was induced by contacting with various metal ions, including Fe^{+3}. The retention capacity of the carrier as well as long term stability of various strains during operation were excellent. It is possible that the binding mechanism of carrageenan to a cell is the formation of partial covalent bonds between the support and metal, and between the same metal and the cell. It is also possible this metal may activate enzymes which govern cell metabolism. Whatever the true mechanism is, addition of inorganic metal ions to organic carriers greatly influences the support retention capacity.

4.1.2   Effect of Support Surface Area on Microbial Coupling

The necessity of porous carriers for enzyme immobilization is well accepted in enzymology because porous materials usually provide more surface area for enzyme loading. Pores of the carrier also protect the enzymes from turbulence in reactors operating at high flow rates. What about microbial loading? Is there any relationship between the amount of mycelium formed inside a porous structure and the amount of surface area available for it? Navarro and Durand[25] investigated the effect of changing the surface area of porous silica beads on yeast immobilization, by varying this parameter upward from 6 m²/gr. The beads were first grafted with γ-aminopropyltrimethoxysilane; then yeast coupling was performed either directly after amino-silane treatment or after amino-silane treatment followed by glutaraldehyde activation. For amino-silane treated silica, maximum bioaccumulation (3mg cells/gr support) was achieved by using a carrier with a surface area of 6 m²/gr. Increasing the surface area of the support to 35 m²/gr sharply lowered the microbial loading, and raising the support surface area further precluded any bioaccumulation at all.

Activating the same carrier with glutaraldehyde yielded considerably different results: Although raising the surface area from 6 m^2/gr to 35 m^2/gr did not permit greater cell loading, the amount of immobilized biomass remained constant at 3 mg/gr support (equal to the level for non-activated silica). Further increasing the support surface area from 50 to 100 m^2/gr caused only a moderate rise in cell loading, but an even higher surface area resulted in sharply greater biomass accumulation, with a maximum of 16 mg/gr support.

The results thus show that after amino-silane treatment, only low surface area silica has affinity for yeasts. Amino-silane followed by glutaraldehyde activation places more reactive groups on the porous silica, permitting a higher retention capacity, as was observed with most porous beads. The authors concluded that increased biomass accumulation resulted from imine bond formation between carrier carbonyl groups and amino groups on the cell surface. Although the surface area calculations are too high, interpretation of results obtained by using supports with larger surfaces is not affected. Clearly, unactivated silica exhibits a rather low retention capacity for *Saccharomyces carlsbergensis* cells.

The effect of support surface area on *P. chrysogenum* and *S. olivochromogenes* bioaccumulation can be roughly calculated from the average pore radius of the carrier,[6] which is inversely proportional to the surface area of the support. Experiments were performed with seven unactivated inorganic carriers (fritted glass, cordierite, zirconia ceramic). Immobilization was carried out by adsorption, and the amount of protein that formed inside the pores of the carrier was estimated after 24 and 48 hours.

Following 48 hours of growth in an appropriate medium, the highest mycelial accumulation of *P. chrysogenum* was observed on four cordierite type carriers, the surface areas of which differed drastically. Moreover, greatest biomass accumulation (2640 γ/gr support) occurred on the spinel-zirconia carrier, which had the largest average pore radius and therefore the lowest surface area compared with other cordierite supports. On the other hand, low biomass accumulation (168 γ/gr support) was observed on the fritted glass carrier, with its rather high surface area. It should be emphasized that the supports were not activated and mycelial accumulation was small.

Unlike *P. chrysogenum*, *S. olivochromogenes* microbes preferred fritted glass over cordierite. However, a drastic surface effect was observed after 24 hours of growth but was completely eliminated after 48 hours of incubation. The amount of mycelium formed at that time was nearly the same on all carriers under investigation, although the total surface area per gram of support varied as much as nine fold.

It is known that for enzyme immobilization, a porous carrier is preferable to a non-porous carrier. The same is true for microbial adsorption, but by contrast with immobilized enzymes, the amount of available surface area does not play any crucial role. Nevertheless, this parameter may contribute to the stability of a biocatalyst during continuous operation if there is partial covalent bond formation between the cell and support. It can also play a role in activating enzymes responsible for cell metabolism.

Although cell adsorption to an inactivated inorganic carrier does not depend on its surface area or porosity, immobilization is greatly influenced by the concentration of reactive groups on the support surface. For activated inorganic carriers, increasing the surface area results in higher microbial loading.

4.2 Summary

A great variety of organic and inorganic carriers are available for microbial immobilization. They differ in the quality and quantity of reactive groups capable of interacting with the cell surface. Although the major contributor to performance of biocatalyst systems is the microbial cell itself, the role of a support should not be underestimated. A prudently chosen carrier may enable cell immobilization in the living state, thus increasing half life of the system, cofactor regeneration, and the rate of product formation.

In spite of the fact that organic carriers have more reactive groups on their surfaces compared with inorganic supports, most recent data show that incorporation of metal into organic matrices increases their retention capacities, activates enzymes governing cell metabolism, and changes the microbial cells' wall permeability.

Until recently, carriers were mainly chosen empirically. However, understanding the mechanisms of microbe-support interactions, together with knowledge of carrier properties and carrier contributions to cell metabolism can result in tailoring specific supports for particular applications. This would lead to higher activity and longer stability of the microbial biocatalysts.

Literature

1. Chibata, I., Tosa, T., *TIBS*, April, 88 (1980)
2. Vijayalakshmi, *et al.*, *Ann. NY Acad. Sci.*, **249** (1979)
3. Vieth, W. R., Venkatasubramanian, K., *ACS Symposium*, Series 106, 1 (1979)
4. Durand G., Novarro, J., *Process Biochemistry*, September, 14 (1978)
5. Marcipar, *et al.*, *Biotech Lett.* **1**, 2, 65 (1980)
6. Messing, R. A., Oppermann, R. A., and Kolot, F. B., *ACS Symposium*, Series 106, 10 (1979)
7. Kolot, F. B., *Developments in Industrial Microbiology*, **21**, 295 (1980)
8. Kolot, F. B., *Process Biochemistry*, Oct./Nov., 2 (1980)
9. Chibata, I., Tosa, T., Tokata, I., US Patent 4,138,261 (1979)
10. Chibata, I., *SCA Symposium*, Series 106, 13 (1979)
11. Toda, K., *Biotech. Bioeng.*, **17**, 487 (1975)
12. Tsao, G. T., Chen, L. F., US Patent 4,063,017 (1977)
13. Johnson, D. E., Ciegler, A., *Archiv. Biochem. Biophys.*, **130**, 384 (1969)

14. Jack, T. R., *Biotech. Bioeng.*, **XIX**, 631 (1977)
15. Vieth, W. R., *et al.*, US Patent 3,972,776 (1976)
16. Vieth, W. R., *et al.*, *Biotech. Bioeng.*, **XV**, 565 (1973)
17. Venkatasubramanian, K., *et al.*, *J. Ferm. Techn.*, **52**, 4, 268 (1974)
18. Rotman, B., *Bacteriological Review*, **24**, 2, 251 (1960)
19. Daniels, S. L., *Developments in Industrial Microbiology*, **13**, 21, 1972.
20. McLaren, A. D., Skrujins, J. J., *Can. Journal of Microbiology*, **13**, 21 (1963)
21. Hattori, T., Furusaka, C., *J. Biochemistry*, **50**, 4, 312 (1961)
22. Hattori, R., *et al.*, *J. of General and Applied Microbiology*, **18**, 272 (1972)
23. Chibata, I., *et al.*, *J. of General and Applied Microbiology*, **27**, 878 (1974)
24. Suzuki, S., Karube, I., *ACS Symposium*, Series 106, 59 (1979)
25. Navarro, J. M., Durand, G., *European J. Applied Microb.*, **4**, 243, 1977
26. Navarro, J. M., *SMT Coloque*, Toulouse, France, 211 (1978)
27. Mole, M., *et al.*, US patent 4,009,286 (1977)
28. Weetall, H., *et al.*, *Biotech. Bioeng.*, **XVI**, 689 (1974)
29. Barker, S. A., *et al.*, US Patent 3,912,593 (1975)
30. Kennedy, J. F., *et al.*, *Nature*, **261**, 242 (1976)
31. Helmstetter, C. E., *J. Mol. Biology*, **24**. 417 (1967)
32. Helmstetter, C. E., Cooper, F., *J. Mol. Biology*, **30**, 507 (1968)
33. Thonart, P., *et al.*, *Enzyme and Microbe Technology*, **4**, 191 (1982)
34. Van Haecht, J., *Biotech. Bioeng.*, **27**, 3, 217 (1985)
35. Adlercreutz, P., *Appl. Micro & Biotech.*, **22**, 1, 1 (1985)
36. Horitsu, H. *et al.*, *Appl. Micro & Biotech.*, **22**, 1, 8 (1985)
37. Schnaar, R. *et al.*, *Anal. Biochem.*, **151**, 2, 268 (1985)
38. Zueva, N., *Prikl. Biokhim.*, **21**, 3, 333 (1985)
39. Adlercreutz, P. *et al.*, *Appl. Micro & Biotech.*, **20**, 5, 296 (1984)
40. Mauz, O. *et al.*, *Ann. NY Acad. Sci.*, **434**, 251 (1984)
41. Krysteva, M. *J. Appl. Biochem.*, **6**, 5–6, 367 (1984)
42. Lobarzewski, J., *Biochem. Biophys. Res. Commun.*, **121**, 1, 220 (1984)
43. Tysiachnaia, I., *Prikl. Biokhim. Mikrobiol.*, **20**, 1, 79 (1984)
44. Kuu, W. *et al.*, *Biotech. Bioeng.*, **25**, 8, 1995 (1983)

5

Applications of Yeast Biocatalysts

The fifth chapter reviews applications for some of the more interesting biocatalysts formed by yeast immobilization: beer fermentation, β-galactosidase enzyme production, sucrose inversion, and phenol degradation. (Ethanol fermentation by immobilized yeasts from different substrates, and the economics of such operations will be described in Chapter 6.) The types of reactors used, half lives, and temperature stabilities of the systems are present.

By contrast with bacterial cells, only a few processes based on attached yeasts are now being researched. However, the properties of yeast cells and their performance in fixed bed reactors indicate that yeast biocatalysts ideally fit the new trends in fermentation technology.

5.0 Beer Production

Beer production is based on fermentation of wort sugars (glucose, maltose, maltotriose, etc.) via the Embden-Meyerhoff-Parnas pathway by *S. cerevisiae* or *S. carlsbergensis* cells. Transformation can be a batch or a continuous flow process. Batch conversion takes from 72 to 96 hours, during which a four to five fold increase in the number of yeast cells occurs. Wort carbohydrates are utilized, and ethanol, CO_2, and minor byproducts are formed. The reaction is exothermic; heat is liberated and the temperature rises.

Another way of carrying this process out is by utilizing attached cells. *S. carlsbergensis* yeasts were immobilized by entrapment into different polymer materials such as polyethylene, cellophane, caprone, and phoroplast. The attached cells were used in a column reactor (7 liter volume) through which wort (pH = 3.5) was pumped. Several important observations were made: the adsorption process was very quick, taking only 30 minutes; cell attachment to the support depended on the type of polymer used, pH of wort, and cation concentration in the solution. Washout of cells from the matrix did not occur even at a dilution rate 30 times higher than that accepted in industry. Carriers retained up to 55–75% of the cells in the matrix, which could be regenerated and reused. The rate of product formation was 24 hours faster compared with batchwise brewing due to a more rapid increase in yeast cell concentration (Kolpackni, Isaeva, 1976).

A new fermentation and maturation process for beer production was developed by Berdelle-Hilge. A bio-reactor, constituted of a layer of yeast (30mm thick) retained on a porous plate (0.04 m^2), was fed with a medium at a flow rate of 1.5–2.0 hours per passage (SV = 0.5–0.67 hr^{-1}) under anaerobic conditions. The subsequent maturation stage (24-48 hours) reduced the diacetyl content of the beer. The finished product exhibited the desired degree of fermentation.

Navarro et al. (1976) studied beer production by S. carlsbergensis cells immobilized via adsorption to polyvinylchloride, kieselguhr, and porous brick. The system consisted of two columns with a heat exchanger installed between them. The medium (12° Plato brewing wort, pH = 5.5) was passed through the first column (25–30°C) at a dilution rate of 250 ml/hr. This reactor fermented wort to beer in 90 minutes. The product was next passed through the heat exchanger, where its temperature fell to 0–5°C, and then through the storage column, in which beer maturation occurred. This process offers several advantages compared with other methods of operation previously discussed: different types of beer can be easily produced (only young beer, or beer with different degrees of maturation, depending on residence time in the storage column), the half-life of the packed bed seems to be unlimited: fermentations have been performed more than eight months without any yeast addition. The reactor can be stopped, stored, and reused at any time. The beer produced had an alcohol content of 3.7–4.2% and a pH from 3.8 to 4.3.

A process for continuous brewing was patented by Moll et al. in 1977. This method is rather similar to the work already described but is somewhat more sophisticated. Boiling, treating, barreling of the wort, fermentation, storage, treatment of fermented liquid as well as filtration, are designed on a continuous flow basis, and all phases of manufacturing are performed under fully sterile conditions. The system utilizes yeast cells attached by adsorption to porous materials such as diatoms, polyvinyl chloride, and plastics. Immobilized microbes are placed into both, the fermentation and storage columns, which work at temperature regimes of 30°C and 0–5°C, respectively. The young beer is passed through a treatment tower containing a protease enzyme such as papain, pepsin, or chymotrypsin fixed to a support (brick or glass). The purpose of the storage and treatment columns is to reduce the amount of diacetyl, decrease the quantity of sulfur compounds, and to impove the quality of product formed. Combining the immobilized microbial cell and immobilized enzyme systems with completely sterile continuous substrate flow makes this process very attractive from the industrial and technological points of view. Most importantly, the final product exhibits all the qualities of beer now commonly brewed by batchwise fermentation in vats. However, synthesis by the immobilized cell reactor is performed much more rapidly and cheaply.

According to a statistic, 102 U.S. breweries produced 160,575,908 barrels (1 barrel = 32.5 gal = 123.0 liters) of beer in 1975. Utilization of attached microbes may represent a drastic change in existing technology.

5.1 β-galactosidase Enzyme Production

The lactase enzyme splits milk lactose to glucose and galactose. Reducing lactose content in dairy foods by enzymatic hydrolysis can improve product quality, sweetness, and can increase carbohydrate solubility, as well as provide a low lactose product for the sugar intolerant part of the population. This process could also be used for treating whey, a byproduct of cheese manufacturing containing up to 5% lactose. It has been estimated that in 1979 alone, 42.4 billion pounds of whey were produced, and half was discarded as waste. Due to new polution laws, such practice cannot continue, thus forcing the cheese manufacturing industry to look for new ways of disposing unwanted whey.

Different immobilized enzyme systems were compared and a great variety of techniques were tested for lactase attachment. Problems associated with enzyme production and concentration, protein adsorption to different support materials and its mechanisms, half lives and temperature stabilities etc. were discussed in great detail. Some of these parameters limit the usefulness of immobilized lactase systems. An alternative to such operations is using attached cells as the enzyme source.

Several microorganisms were screened as potential producers of β-galacto-sidase. The *Saccharomyces lactis* strain was selected for further study because it hydrolyzed up to 47% of the lactose. Microbe immobilization by entrapment into acrylamide gel was affected by the cell concentration, the amount of acrylamide monomer, and polymerization time. The best results were achieved when the cell suspension contained less than three grams of cells in 10 ml of phosphate buffer, using a 50% acrylamide solution and a polymerization time of only 15 minutes. This method of attachment resulted in microbes retaining 61% of their β-galactosidase activity vs. 29% for immobilized enzymes. Entrapped cells were wrapped by a nylon cloth impermeable to matrix beads and tested repeatedly in batchwise operations with a 4.5% lactose solution (pH = 6.5) heated to 30°C.

After eight repeated runs, almost 100% of the initial enzyme productivity remained, which strongly points to the possibility of maintaining high β-galactosidase activity by entrapped cells. Immobilization did not change the pH profile or thermostability of attached microbes compared with free cells and the immobilized enzyme. Also, K_m value for the enzyme inside attached cells was almost the same as K_m for the free enzyme. Deproteinized filtrates of skim milk were treated with an entrapped yeast preparation, and the changes in milk flavor, sweetness, and lactose hydrolysis were observed. β-galactosidase activity was also confirmed by the disappearance of lactose and presence of spots of glucose and galactose on a thin-layer chromatogram. Unfortunately, this system was not adapted to continuous column operation, but the data clearly suggest the feasibility of such a process.

5.2 Sucrose Inversion

Sucrose inversion sweetens sugar products. The reaction can be carried out as an acid, enzyme, or immobilized microbe process. The disadvantage of acid inversion is formation of undesirable oligosaccharides. In current industrial practice, the invertase enzyme is immobilized to strong cationic resin, and used in continuous column processes. However, the biocatalyst suffers from substrate inhibition. Transformation by living immobilized microbes is thus an interesting approach.

The immobilized cells were placed into a continuously fed fluidized bed reactor aerated by air bubbles. A 58.4mM solution of sucrose in buffer (pH = 5.0) was passed through the reactor at 47.5°C. Seventy percent of sucrose hydrolysis was achieved under these conditions, and this rate of inversion was stable for about 100 hours of operation. Conversion reached steady state at air flow velocities higher than 6cm/min. The reaction rate in the system depended on carrier radius, interparticle concentration of enzyme, and external concentration of substrate.

5.3 Phenol Degradation

A multistep reaction involving four enzymes and based on decomposition of an aromatic compound was carried out by *Candida tropicalis* cells immobilized via entrapment into an ionic polymer. Enzyme activity within the cells was induced either by repeated addition of phenol during yeast growth or by isolating yeast colonies on a medium containing phenol as the sole carbon source.

Table 1 Processes Based on Immobilized Yeasts*

Microorganism	Method of Immobilization	Substrate	Product	Half Life	Temperature	% Conversion	References
S. carlsbergensis	Entrapment	Wort	Beer		25		Appl. Biochem. Microb., 1976
	Adsorption	Wort	Beer	8 mo.	25		Industries alim., 1976 & U.S. Patent 4,009,286
S. cerevisiae	Adsorption	Glucose	Ethanol	1 mo.	30	46	Navarro, 1978
S. carlsbergensis	Entrapment	Glucose	Ethanol	3 mo.	30	100	Chibata, 1979
S. cerevisiae	Entrapment (co-imm. with enzyme)	Cellobiose	Ethanol	20 days	22	80	Hagerdal, 1979
S. lactis	Entrapment	Lactose	Glucose & galactose	Repeated batch ferm.	30		Ohmiya, 1977
S. pastorianus	Entrapment	Sucrose	D-Glucose & D-fructose	5 days	48	70	Tosa, 1975
Candida tropicalis	Entrapment	Phenol	CO_2 & H_2O	19 days	30	100	Klein, 1979

*Fixed bed reactors

Cells were then harvested, washed, and immobilized; spherical or elipsoidal particles with mean diameters from 3 to 5mm were used in a packed bed circulation column reactor.

A phenol solution (3mM) was saturated with pure oxygen, an essential co-substrate for this transformation, and continuously fed to the reactor. The operational stability of the system was found to be 19 days. Catalytic activity of entrapped microbes was raised by reincubating cells in a nutrient medium. The authors observed that reincubation resulted in the yeast cell concentration increasing by a factor of 11 inside the polymer. However, the higher cell count lead to a greater transport limitation, causing the absolute phenol consumption rate to rise by only a factor of 2.5.

Literature

1. Doran, P. et al., Biotech. Bioeng., **28**, 1, 73 (1986)
2. Mozes, N., Rouxhet, P., Appl. Micro. & Biotech., **22**, 2, 92 (1985)
3. Jain, V. et al., Biotech. Bioeng. **27**, 3, 273 (1985)
4. Cantarella, M., Migliaresi, C., Appl. Micro & Biotech., **20**, 4, 238 (1984)
5. Jones, C., White, E., Yang, R., Ann. NY Acad. Sci., **434**, 119 (1984)
6. Chun, Y. et al., J. Gen. Appl. Micro., **27**, 505 (1981)
7. Kolot, F., Dev. in Ind. Micro., Vol. 21 (1980)
8. Chibata, I., US Patent 4,138,292 (1979)
9. Chibata, I., ACS Symposium, Series 106, 187 (1979)
10. Chibata, I., Inter-society US-Japan Microb. Congress. Honolulu, Hawaii. Abstract, p. 202 (1979)
11. Hogerdal, B., Mosbach, K., International Congress on Eng. and Food. Helsinki, Finland (1979)
12. Klein, J., Hackel, U., Wagner, F., ACS Symposium, Series 106, 101 (1979)
13. Neujahr, H., Prog. Biochem., **13**, 6, 3 (1978)
14. Tsao, G., Prog. Biochem., October (1978)
15. Dickensheets, P. et al., Biotech. Bioeng., **19**, 365 (1977)
16. Kierstan, M., Burke, L., Biotech. Bioneg., **19**, 387 (1977)
17. Moll, M., Durand, G., Blachere, H., US Patent 4,009,286 (1977)
18. Navarro, J., Durand, G., Eur. J. Appl. Micro., **4**, 24 (1977)
19. Kolpackni, A., Isaeva, V., Appl. Biochem. & Micro., **12**, 6, 866 (1976)
20. Kennedy, F. et al., Nature, **261**, 242 (1976)

6

Microbial Catalysts
for Solvent Production

The purpose of this chapter is to present the most recent trend in solvent fermentation, application of immobilized living microbial cells for production of industrially important solutants: ethanol, acetone-butanol, 2,3 butanediol, and n-butanol-isopropanol.

Various systems for ethanol biosynthesis are compared and advantages of immobilized bacterial systems over immobilized yeasts are presented. Acetone-butanol formation in repeated batch and continuous column operations by immobilized spores of *Clostridium acetobutylicum* and by immobilized vegetable cells (cells in the logarithmic phase of growth), as well as continuous butanediol production by entrapped *Enterobacter aerogenes* cells are described. Butanol-isopropanol fermentation in a column reactor by entrapped *Clostridium butylicum* cells is also given.

Methods of cell immobilization, their mechanisms, the effect of cell concentration inside the matrix, cell activation in the immobilized state, operational stability, product yield, and advantages and limitations of the immobilized cell systems compared with free cells, are all discussed.

6.0 Ethanol Fermentation

Ethanol can be produced either by chemical synthesis (direct hydration of ethylene) or via fermentation. In 1979, 196 million gallons of ethanol were made in the USA by chemical synthesis, and mostly used as solvent (55%) and chemical intermediate (35%). Although plant capacity could have been increased to 270 million gallons, the production rate remained at the above level during 1978–1980. Despite the fact that a blend of gasoline (90%) and ethanol (10%), (gasohol), can be used as fuel, neither chemical nor established fermentation industries are willing to make ethanol for gasohol purposes. To encourage industrialists, the U.S. administration is providing loans and tax credits for ethanol synthesis from renewable sources.

According to the U.S. gasohol program, by the end of 1981, 500 million gallons of ethanol were produced, by 1985 the production level was targeted for 1.8 billion, and by 1990 for 9.0 billion gallons. The new trends in ethanol fermentation technology, which will be discussed, can make a substantial contribution to the success of the ongoing program.

Alcohol can be produced from different fermentable substrates such as grain sugars, molasses, grape juice, glucose, lactose, whey (a waste product of cheese manufacturing), etc. by *Saccharomyces cerevisiae, S. carlsbergensis, S. sake, Kleyveromyces fragilis* yeasts, or bacterial strains (*Z. mobilis*). The biochemical reactions of ethanol synthesis from glucose are well known. The process is a multi-step enzymatic reaction catalyzed by various enzymes located in microbial cells and requires regeneration of ATP and coenzymes. It is an anaerobic process, but a trace of oxygen is required for biosynthesis. Below a certain level, yeast cells become oxygen starved and ethanol production decreases. At a high oxygen level, a shift from an anaerobic to an aerobic process occurs, cell mass increases, and less product is formed. Ethanol can be produced batch wise or on a continuous flow basis. Formation is limited by three factors: end product inhibition, low cell concentration, and substrate inhibition.

To overcome ethanol inhibition, the alcohol must be constantly removed from the fermentor. This can be achieved by performing the fermentation on a continuous basis under a sufficient vacuum to boil off ethanol as it is produced. To increase the low cell concentration, cell recycling is used: A portion of the microbes is separated from the broth and returned to the fermentation vessel. To reduce substrate inhibition, the following approaches may be considered: 1. microbial mutation followed by screening for survivals to utilize high sugar concentrations, 2. stepwise adaptation of yeasts to high substrate concentrations. The combination of yeast recycling with a continuous vacuum process results in high product yield. A comparison of different fermentation systems tested for alcohol production is presented in Table 1.

The optimal sugar concentration for free yeast (batch, continuous, continuous with recycling) and free bacterial systems (continuous with recycling) was 10%. But only free yeasts (vacuum with recycling) could utilize a sugar concentration of 50%. The large cell density (124 gr/l) and high ethanol production (82.0 gr/l/hr) point to a good potential for the vacuum process based

Table 1   Systems Tested for Ethanol Production

System	Methods of Immobilization	Sugar Conc. %	Cell Mass Concentration g/l/No./ml	Maximum Ethanol Production g/l/h	Temperature	Residence Time (h)	Half Life	Percent of Theoretical Yield	Reference
Free yeasts									
A) Batch		10	5.6/	2.2	35				1
B) Continuous		10	10–12/	7.0	35				1
C) Continuous with recycling		10	50/	29.0	35				1
D) Vacuum with recycling		50	124/	82.0	35				1
Immobilized yeasts	Entrapment	10	5.4×10^9	50–100	30		3 mo	100	2
		15							3
		20							3
		25	10^{10}	45.6	30		80 days	96	3
Immobilized yeasts	Co-immobilized with an enzyme	5.0		20.0	22	2.5	20 days	80	4
Immobilized yeasts	Adsorption	10.0		120.0	30		1 mo	46	5
Free bacteria (continuous with recycling)		10.0	38/	120.0	30				6, 7
Immobilized bacteria	Entrapment	15.0	26/	53	30	1.2	1 mo		8–10
		17.5	/10^8	68	30		1 mo		

on free yeasts. However, a continuous process with recycling based on free bacterial (Z. *mobilis*) cells provided an even better product yield (120 gr/l/hr).

Operations based on immobilized growing yeasts and bacterial cells are also very attractive, since they can be performed at rather high substrate concentrations (17.5%–25%) and provide high ethanol yields: 50–100 gr/l/hr for immobilized yeasts and 68 gr/l/hr for immobilized bacteria. Residence times in the immobilized cell reactors are rather short (2.5 hr for yeasts, 1.2 hr for bacteria), and the systems are stable during continuous operation for one to three months.

Thus, the four systems: free yeasts (vacuum process with recycling), free bacteria (continuous process with recycling), immobilized yeasts (via entrapment), and immobilized bacteria (via entrapment) will probably be the processes of the future.

Application of immobilized microbes in continuous column reactors is the most modern approach in alcohol production technology. It combines several advantages of the systems already discussed. Namely, a high cell concentration can be obtained, the reaction rate is accelerated, and operation can be performed at a large dilution rate with little washout of cells from the support. End product inhibition is also eliminated because alcohol is constantly removed; anaerobic conditions can be achieved since cells are entrapped into the gel matrix, and by contrast with already described processes, costly equipment design of fermentors, agitators, etc. is eliminated.

Several research groups in different parts of the world are currently working on alcohol production using immobilized yeast and bacterial cells. Each group uses completely different techniques for microbial attachment. The investigations conducted in France, Japan, Sweden, and Australia will be compared and discussed.

In 1976, Navarro et al.[5] studied continuous ethanol production in a column type, 50×5 cm reactor, using the S. cerevisiae strain STV 89 immobilized by adsorption to brick and polyvinyl chloride chips. The microbial catalyst was placed into the column, and a medium containing glucose, malt extract, and salts was recirculated through the reactor at 30°C every seven hours. Maximum alcohol production of 120 gr/l occurred after two days of incubation; at this time, 46% conversion of sugar to alcohol was achieved. The column operated continuously for 30 days without addition of new microorganisms, leading the author to conclude that good performance was due to a high concentration of cells retained in the porous material.

Since 1979, ethanol production in Japan has been studied using yeast cells entrapped into K-carrageenan gel. The polysaccharide was dissolved in saline (forming solutions with concentrations ranging from 0.05% to 20%) and heated to 37–50°C. A yeast suspension was also warmed to the same temperature and added to the carrageenan solution.[2] The mixture was cooled and contacted with gel inducing agents; finally, the matrix was contacted with gel-hardening agents and granulated into bead-like particles in which the cell concentration was estimated at 3.5×10^6 cells/ml. The yeast cells thus immobilized were alive to multiply inside the support during incubation in a complete medium containing carbon/nitrogen sources, and salts (pH = 5.0). After 60 hours of incubation, the concentration of living cells increased 1500 fold, reaching 5.4×10^9 cells/ml of gel. The carrier particles (mean diameter 4mm, 20 ml column volume) were packed into a jacketed glass column reactor (12.5×2.5 cm), through which a complete medium was pumped at a flow rate of 20 ml/hr and 30°C (retention time of 1 hour). Some microbes washed out of the matrix but the number of cells in the solution remained at a level of 10^6-10^7 cells/ml, much lower than the microbe concentration inside the gel of 10^9 cells/ml. This led researchers to conclude that the free microbes in the medium were detached from the multiplying colony on the support surface. A steady state in the number of cells and ethanol production were achieved after five to six days of operation. Continuous alcohol synthesis at the level of 50–100 mg/ml was observed for a thirty day period. Conversion of glucose to ethanol reached 100% of the theoretical yield even at the fast flow rates such as a retention time of one hour. Therefore, this system appears to be very efficient and promising from the industrial standpoint.

6.0.1 Continuous Ethanol Production by Immobilized Cells at High Substrate Concentrations[3]

As mentioned, substrate inhibition is one of the inherited problems in ethanol production by the free cell system. This disadvantage is reduced in the immobilized cell column.

Continuous ethanol formation by attached cells at high substrate concentrations (e.g. 25%) makes this system even more attractive.

Eleven immobilized growing cell preparations were formed by microbial entrapment into K-carrageenan gel. The catalysts were screened for highest ethanol yield using two glucose concentrations, 10% and 25%. The preparation formed by entrapment of Saccharomyces cerevisiae IFO 2363 cells exhibited the highest product yield (34.8 mg ethanol/ml of gel/hr) at a 25% substrate concentration compared with the ten other strains.

A fresh preparation of catalyst was formed by immobilizing S. cerevisiae IFO 2363 cells into K-carrageenan at an initial concentration of 9×10^6 cells/ml. Equal amounts of beads (20 ml) were then packed into four columns and complete media of varying glucose concentrations (10%, 15%, 20%, 25%) were pumped through the reactors for 70 hrs at a flow rate of 25 ml/hr. The largest microbe concentration inside the gel (5×10^8 cells/ml) occurred at the lowest glucose concentration of 10%. The highest glucose concentration in the feed gave a catalyst with the smallest cell count (9×10^7 cells/ml).

These results strongly suggest substrate inhibition. Based on that observation, the following strategy was adopted: Gels with low cell counts ($\sim 10^7$ cells/ml gel) were packed into columns through which a complete medium with 10% glucose was pumped for six days at the flow rate of 25 ml/hr. Consequently, the cell concentration increased by three logs (from 10^7 to 10^{10} cells/ml gel). After cell density reached a steady state of 10^{10} cells/ml gel, substrate was fed to the reactors at concentrations increasing every six days (15%, 20%, 25%). Retention times for the different substrate concentrations were 0.8, 1.6, 2.2, and 2.5 hours, respectively (Fig. 1).

At the highest glucose concentration (25%) and a retention time of 2.5 hours, 114 mg/ml of ethanol were produced. The transformation of sugar to alcohol was 96% of theoretical activity and the conversion efficiency, expressed as grams of product per gram substrate was 45.6%.

Thus, the system described is quite efficient for ethanol synthesis at high glucose concentrations.

6.0.2   Continuous Ethanol Production by Free Bacteria[6-7]

Alcohol production by free *Zymomonas mobilis* bacteria in the cell-recycling system has several advantages over free yeasts (using the same process). These are: 1. higher ethanol yield, 2. production at higher glucose concentrations, 3. greater ethanol tolerance, 4. bacteria produce less biomass than yeasts.

Z. mobilis cells utilize glucose via the Entner-Doudoroff pathway, forming ethanol and one mole of ATP per mole of substrate, whereas yeasts metabolize glucose anaerobically via the glycolytic pathway, giving ethanol and two moles of ATP per mole of glucose. The ethanol yield for free bacteria in a continuous process with recycling was 120 gr/l/hr, but only 29.0 gr/l/hr for free yeasts in the same process (Table 1).

However, since yeasts are continuously washed out, they can be collected and partially sold as single cell protein, which gives "yeast credit" to ethanol producers. Ethanol producing systems based on bacteria cannot get such credit, but because *Z. mobilis* strains form less biomass compared with yeasts, and ethanol synthesis by bacteria at high flow rates is greater, the expenditure on waste treatment should not discourage producers.

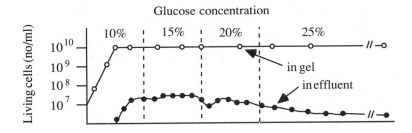

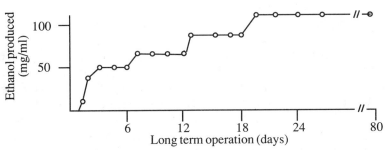

Figure 1 Long term ethanol production in immobilized yeast reactor (stepwise feeding system).[3]

6.0.3 Continuous Ethanol Production by Immobilized Bacteria[8-10]

Bacterial cells were immobilized by entrapment into Ca-alginate gel, carrageenan gel, adsorption to borosilicate glass fiber pads, and via flocculation. Column type reactors of various sizes, fixed bed (constant volume), and flocculent bed (variable volume) were used in the study.

Maximum productivity of the reactor with immobilized *Z. mobilis* cells using

15% glucose was 53 gr/l/hr at a dilution rate of 0.8 hr^{-1}. The average cell concentration reached 26 gr/l and the residence time was only 1.2 hours.

Slightly greater ethanol levels and less biomass washout were observed at the same high flow rate for carrageenan entrapped bacteria compared with Ca-alginate entrapped cells.

Bacterial accumulation occurred in both types of gels with extended operation, but destruction of the matrix structure was apparent only for the Ca-alginate support.

The specific rate of glucose intake and ethanol production by immobilized Z. mobilis cells was two times lower compared with free cells. This reflects a diffusion limitation, which is an inherent problem in the immobilized cell reactor.

Ethanol synthesis at a high substrate concentration of 17.5% was studied using the Z. mobilis ATCC 10988 strain entrapped into K-carrageenan gel. Performance of immobilized bacteria was compared with immobilized yeasts after the latter were attached by entrapment into an identical matrix. The beads (170 ml) were packed into 225 ml reactors and continuously supplied with a substrate medium containing glucose (17.5%), KH_2PO_4 (0.1%), NH_4Cl (0.2%), and yeast extract (0.3%).

Although both systems performed similarly for 5–11 days at the low flow rates of 13–25 ml/hr, a drastic difference in reactor activity appeared at high flow rates of 75–150 ml/hr, and over longer runs (17–30 days). In the bacterial reactor, raising the medium velocity from 30 to 150 ml/hr increased ethanol formation to 70–80 gr/l, leaving no residual sugar in the effluent. By contrast, raising the flow rate of the yeast reactor by the same amount caused a steady decline in ethanol production (mg/ml gel/hr) and lowered the alcohol concentration in the effluent (gr/l), which finally fell to only half of the bacterial level (40 gr/l).

Obviously, the immobilized bacterial system is more efficient than the immobilized yeast reactor under similar conditions, and therefore has excellent potential (Fig. 2).

The authors proved that Z. mobilis cells were alive inside the matrix and hence represented "immobilized growing bacterial cells." The cell concentration increased after placing the support into a complete medium containing glucose, phosphate, ammonium, and yeast extract. When the solution was limited in any one of the above components, the number of cells in the matrix decreased and ethanol productivity also declined.

It was thus demonstrated that of the numerous processes compared in Table 1, only four are most interesting from the industrial stand point: the vacuum process with recycling for free yeasts, continuous process with recycling for free bacteria, immobilized yeasts, and immobilized bacteria. Apparently, none

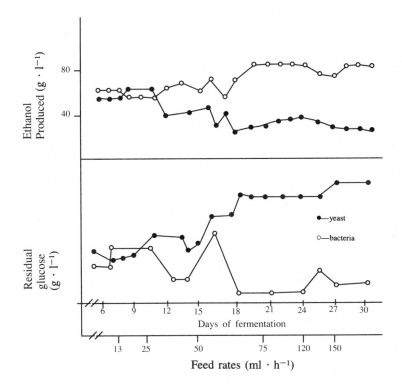

Figure 2 Comparison of ethanol production by immobilized yeasts and immobilized bacteria (17.5% glucose concentration and varying flow rates).[10]

of the authors used stable mutants in these systems. Application of specifically selected mutants will result in more effective immobilized cell reactors for ethanol fermentation.

6.0.4 Ethanol Production from Cellulosic Wastes[15–31]

Cellulosic waste, a renewable waste resource, is produced world-wide in excess of two billion tons annually, of which 1010 million tons are available in the United States alone. On average it consists of three components: cellulose, hemicellulose, and lignin, in a 4:3:3 ratio. Cellulose is a polymer of glucose; hemicellulose is a polymer of pentoses (xylose and arabinose) and hexoses (mannose); lignin is a polyphenolic compound. Hemicellulose can be much more easily hydrolyzed than cellulose, which consists of an amorphous, easily hydrolyzed part (15%), and a crystalline part (85%) surrounded by lignin. Thus, the first step in treating cellulosic waste is breaking the crystalline structure of cellulose, converting it to a soluble form acceptable for acid or enzymatic hydrolysis.

Some cellulosic wastes can be more easily transformed to a soluble form compared with others. For example, agricultural wastes such as cornstalk, bagassa, alfalfa, orchard, and grass consist of up to 30% amorphous cellulose. Pretreatment with the solvent cadoxen results in selective extraction of amorphous and crystalline cellulose, and its separation from lignin, a solid residue. The lignin could then be removed by filtration and used for making phenol, phenolic resins, and as fuel for the plant. Pretreating corn residue and bagassa with cadoxen for 20 hours solvates 85–95% of the soluble cellulose. The solution can be easily hydrolysed to glucose by acid and enzymatic methods.

For other wastes such as newsprint, treatment by weak or strong acids at high temperatures and pressures proved strong enough to convert cellulose to a soluble form.[5–7] Hydrolysis of newprint in a weak acid (1%) at 230°C and 500 psi resulted in a mixture of cellulose (61%), hemicellulose (16%), and lignin (21%).[7] Cellulose degradation in strong acids reduced total sugar yield due to simultaneous glucose degradation, and increased the process cost. Furthermore, since an acid is a non-specific catalyst, it attacked hemicellulose as well as cellulose, forming soluble oligosaccharides, which were quickly hydrolyzed to monosaccharides. The resulting purity of glucose and xylose produced via acid hydrolysis was lower than by an enzymatic process, due to decomposition of the sugars and formation of acetic acid.

In the enzymatic process,[5,6,8] endo-glucanases randomly attack the 1,4-β-linkages of the cellulose chain, while exoglucanases split off cellobiose or glucose units from the nonreducing end of cellulose. Both enzymes are, however, subject to cellobiose inhibition. Also, the β-glucosidase enzyme, which hydrolyses cellobiose to glucose, is inhibited by the glucose. Finally, glucose can be converted to ethanol via a multistep enzymatic reaction, but the product inhibits endo(exo)-glucanases, β-glucosidase, and yeast cell growth. Thus, each step in the transformation of cellulosic waste to ethanol is subject to intermediate and final product inhibition (Fig. 3).

Numerous articles dealing with hydrolysis describe various chemical digesters, methods of substrate pretreatment, as well as the enzymes and regulatory mechanisms involved in biochemical reactions. However, the competition between chemical and enzymatic methods for (hemi) cellulose degradation is of secondary importance to improvement in the fermentation

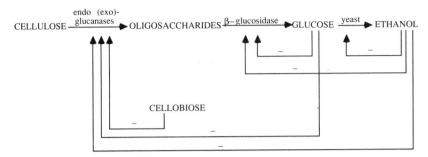

Figure 3.

step. Overall, conversion of cellulosic waste to ethanol is profitable only when:
1. Fermentation proceeds on a continuous basis with a short residence time for the glucose and xylose substrates.
2. Ethanol yield is not less than 6–12%.
3. Ethanol recovery is simple and energy efficient.

Combining waste saccharification, yeast fermentation, and ethanol recovery into one continuous process is very advantageous for fullfilling the above three requirements.

The purpose of this section is to discuss the new approach: a continuous process for converting cellulosic waste to ethanol based on a combination of enzyme-membrane and enzyme-microbe reactors installed in a substrate stream to prevent product inhibition.

The system described includes four separate steps:
1. Substrate hydrolysis via an enzyme-membrane reactor.
2. Concentration of cellobiose and glucose (xylose) via reverse osmosis.
3. Simultaneous conversion of cellobiose to glucose and glucose (xylose) to ethanol in an enzyme-microbe reactor.
4. Ethanol concentration.

Enzymatic hydrolysis of cellulose by cellulase was performed in an Abcor-type ultrafiltration unit containing the DDS GR 6P membrane, which separated the compounds with low molecular weights (20,000MW). Cellobiose and glucose, products of the enzymatic reaction, were therefore divided from the substrate, eliminating inhibition of endo and exo-glucanases by the products.

A high dilution rate (0.28 hr^{-1}) was clearly preferable during enzymatic hydrolysis of cellulose in the membrane reactor since the speed of hydrolysis rose with increasing flow rate. To concentrate the dilute product formed, the authors installed the cellulose acetate membrane DDS-575 into the medium stream, increasing the sugar concentration to 14% by reverse osmosis.

The sugar solution was then passed through the immobilized microbe reactor containing *S. cerevisiae* cells coentrapped with the β-glucosidase enzyme into a sodium alginate support. The enzyme split cellobiose to glucose, while yeasts converted glucose to ethanol. This arrangement eliminated ethanol inhibition of yeast growth because the product was continuously removed from the system; glucose inhibition of the β-glucosidase enzyme was also eliminated since yeasts fully converted glucose to ethanol.

The authors proposed to use electrodialysis and reverse osmosis prior to final product distillation. Electrodialysis eliminates components responsible for reduction in the efficiency of reverse osmosis (RO). Installing the RO unit, which contains the methylated polyacrylamide membrane (IR CHA, France), increased the ethanol concentration to 15–20%. The reverse osmosis and electrodialysis units are probably too costly for ethanol recovery from cellulosic waste, but putting in an extractive bioconversion unit directly after the immobilized enzyme-microbe column seems very advantageous.

A.   Extractive Bioconversion of Cellulose to Ethanol

Another approach to cellulose transformation is applying an extractive bioconversion technique using a two phase system. The enzyme and substrate are confined in one phase, while the product is distributed in the second. The two phase system consists of dextran T-40 and Carbowax PEG 8000.

Enzymatic saccharification of cellulose (Avicel PH-105) by cellulase was carried out with this method utilizing 6.0% dextran and 7.5% carbowax. Extractive bioconversion in this system was better compared with the buffer system, judging by the amount of soluble sugars formed.

Fermentation of ethanol was also carried out in the two-phase and buffer systems, with glucose added to both at various concentrations. The rates of product formation were similar for both methods (50–60%); furthermore, cell enzymatic activity did not decline at the higher sugar concentrations in the two-phase system.

Bioconversion of cellulose (Avicel PH-105) to ethanol using cellulase, β-glucosidase, and *S. cerevisiae* cells was performed in the phase system and compared with the buffer experiment. It was observed that the first system performed better than the second, giving 60–70% conversion after 15 days of operation.

Thus, the results presented above clearly indicate that continuous extractive bioconversion of cellulose to ethanol is a feasible transformation using the two-phase system.

The processes described so far have been based on the assumption that only the hexoses formed during cellulosic waste hydrolysis are converted to ethanol.

Pentoses created during hydrolysis of hemicellulose were not considered. In fact, xylose is not fermented to ethanol by either *S. cerevisiae* or *Z. mobilis*, causing approximately 50 million tons of crop residue, which contains from 10 to 25% xylose, to be wasted each year, and significantly reducing the total ethanol yield from the crop. If pentoses were directly converted to ethanol, the efficiency of alcohol production would increase substantially.

Recently rediscovered *Pachysolen tannophilis* yeasts can ferment xylose directly to ethanol with a high yield. Yeast growth and activity were inhibited by the substrate and product, but this adverse reaction could have been avoided if fermentation were performed in a membrane reactor on a continuous basis. Applying the following three membrane reactors in sequence would result in a much more efficient cellulosic waste transformation process:

1. Immobilized *P. tannophilis* for xylose to ethanol conversion.
2. *Termomonospora thermophilicum* for cellulose to cellobiose and glucose conversion.
3. *Clostridium* (or *Zymomonas mobilis*, *S. cerevisiae*) co-immobilized with the β-glucosidase enzyme into one matrix for cellobiose to glucose and glucose to ethanol conversion.

Nearly unlimited possibilities exist for improving each step in continuous production of ethanol from cellulosic waste. Depending on the specific resource, modifications in substrate pretreatment can be made. Also, separate parts of the system already described can be rearranged in different flow sequences. For newsprint cellulosic wastes, for example, a membrane reactor with *Pachysolen tannophilis* cells can be installed directly after the acid hydrolysis unit and prior to cellulose saccharification. It would also be possible to perform xylose fermentation in a membrane reactor with a two-phase bioextractive system and direct the product to an ethanol evaporation (concentration) unit. Saccharification of cellulose with the cellulase enzyme, splitting cellobiose to glucose by β-glucosidase, and glucose to ethanol bioconversion using bacteria or yeasts should be performed in one sectional membrane reactor, with each part separated by different membranes.

Developing thermophilic and stable mutants with high cellulase activity is necessary. This would enable applying mixed cultures of thermophilic *Termomonospora* for cellulose saccharification, thermophilic *Clostridium* (or *Z. mobilis*, *S. cerevisiae*) for glucose to ethanol conversion, and *P. tannophilis* for xylose fermentation. One should remember that using immobilized bacterial cells is advantageous to utilizing immobilized yeasts for alcohol production because a bacterial catalyst can operate at greater dilution rates with high yield. In each system, a study of various membranes designed to prevent substrate inhibition and increase product concentration before evaporation is necessary.

Another approach to efficient ethanol synthesis is transferring plasmids from the *P. tannophilis* strain to *Z. mobilis*, enabling the genetically engineered microbe to transform pentoses as well as hexoses directly to ethanol.

From the wealth of literature available, it is clear that co-immobilizing an enzyme with one or more microbes inside a single matrix, together with the most recent advances in membrane technology and genetic engineering, can result in development of efficient, specifically tailored continuous processes for ethanol production from cellulosic wastes.

6.0.5   Economics of Ethanol Production

A comparison of total production costs for different fermentation systems is presented in Table 2. As evidenced, sugar dominates the expense of alcohol synthesis and thus dictates the selling price. From this perspective, alternate substrates will have a major effect on the total product cost, possibly reducing it dramatically. For example, alcohol formation by *Kleyveromyces fragilis* cells from cheese whey, now a polutant and a waste product of cheese manufacturing, could be very cost-efficient.

On the other hand, the expense of a continuous process utilizing hydrolysed cellulosic waste materials as substrate is almost the same as of a continuous process with recycling, or for the vacuum process using molasses in the feed, which is due to the relatively high cost of acid hydrolysis. By contrast, the immobilized cell systems are attractive because of their high product yield and low equipment expenditure.

Only currently accepted technology, namely free cell batch wise ethanol production, seems to perform poorly compared with other systems. One should notice, however, that in all types of fermentation the net product cost is lower compared to synthetic alcohol.

A comparison of return on investment for different processes is provided in Table 3. The lowest percent of return (ROI) is calculated for the batch operation. This method cannot presently compete with others and should not attract producers. Continuous fermentation with or without recycling and the vacuum process have ROI ranges from 55.5% to 81.5% if yeast credit is taken into account and from 27.9% to 64.3% without yeast credit. The latter ROI figures are probably more realistic since increasing alcohol production for gasohol purposes would saturate the market for yeast, reducing their cost. Even with a lower ROI, profitable production can be achieved. Gasohol is now not as competitive with gasoline as it was several years ago, but it will become more so as the price of crude oil rises.

The new trends in ethanol fermentation technology, described here, may help the gasohol industry and could decrease our dependence on foreign oil.

Table 2 Ethanol Production Costs for Different Modes of Operation[1]* (cents/gallon)

Ethanol Production Cost	Batch[†]	Continuous[†]	Continuous[†] with Recycling	Vacuum[†] with Recycling	Continuous[‡] Using Cellulose	Immobilized Cells
			Modes of Operation			
Fermentation	16.8	6.6	5.0	4.8	30–40	6.6
Ethanol recovery	8.1	8.1	8.1	6.7	1.0	8.1
Yeast recovery	2.0	2.0	2.4	1.0	0.2	2.0
Storage	0.6	0.6	0.6	0.2		0.6
Sugar	73.7	73.7	73.7	73.7	60.0	73.7
Total product cost	101.2	91.0	89.8	86.4	91.2–101.2	91.0
Yeast credit	—	—	13.3	6.2	13.3	13.3
Net product cost	101.2	91.0	76.5	80.2	77.9–87.9	77.7

*Production cost of chemically synthesized ethanol $1.53–1.65[14a]
†Plant capacity: 78 × 10³ gal/day (95% ethanol); Carbon source: Molasses[1]
‡Calculated for plant capacity of 1 ton/day[14b]

Table 3   Comparison of Return on Investment
for Various Fermentation Processes[1]*

| | Percent Return on Investment | |
Fermentation Process	Yeast Credit	No Yeast Credit
Batch	18.5	3.3
Continuous	55.5	27.9
Continuous with cell recycling	69.7	36.5
Vacuum with cell recycling	81.5	64.3

*Ethanol—95%; selling price—$1.10/gal.

Percent return on investment—(yearly profit/total capital investment) × 100

6.1   2,3 Butanediol Production by Living Immobilized *Enterobacter* Cells[11]

Butanediol, now commercially derived from petroleum, can be easily converted to butanediene, the raw material used in making synthetic rubber. However, high oil prices had intensified the search for alternate sources of petrochemicals. The traditional fermentation process, which has not been used during the last thirty years, is now under reevaluation. Applying immobilized living microbial cells for butanediol synthesis was also recently investigated.

Butanediol production by attached *Enterobacter aerogenes* IAM 1133 cells was described and compared with free cell fermentation. Microbial immobilization was accomplished by entrapment into K-carrageenan gel. Upon entrapment, the gel beads were placed into a complete nutrient medium, causing the cell concentration inside the gel to increase 200 fold (27 mg/gr gel) after 12–24 hours of incubation. Microbial catalyst particles (3–4 mm diameter) were used for substrate conversion in batch (glucose 5%) and continuous operations. In the latter case, the substrate concentration was reduced two fold (glucose 2.5%).

The rate of batchwise 2,3-butanediol production by immobilized cells increased linearly for 30 hours, reaching 4–5 mg/hr/gr gel. The activity stabilized at this level for a ten day period. Entrapped microbes formed six times less of the by product acetone (0.3 mg/ml) than free cells (1.8 mg/ml). Free cells were more sensitive to pH, temperature, dissolved oxygen and glucose concentrations compared with immobilized microbes, which suggests greater stability as a result of cell bonding to the carrier. Entrapped cells were stable at 4°C, but synthesis drastically declined after temperature elevations, as with free cells. Raising the substrate concentration from 5% to 10% caused no increase in product yield of 0.6 mole product/mole substrate.

Continuous butanediol production by the *E. aerogenes* biocatalyst was accomplished after microbe reactivation inside the used matrix. Reactivation was performed in 24 hour intervals by placing the carrier into a nutrient medium. The cell concentration inside the support rose drastically, which was confirmed by dry weight determination and microscopic observation. The rate of product formation by entrapped cells rose parallel to the increasing cell mass inside the gel.

The reactivated immobilized *E. aerogenes* microbes (20gr) were placed into a continuous reactor (working volume-125ml; medium volume-100ml). A medium was pumped through the column at a flow rate of 25 ml/hr (30°C) and slowly agitated by a magnetic stirrer. 2,3-butanediol production remained at the same level (3 mg/ml) for ten days of operation.

The calculated gel activity was 3.7 mg/hr/gr gel, giving a constant yield of 0.5 mole product per mole substrate (Fig. 4).

Although synthesis by entrapped microbes was somewhat lower compared to free cells (1% vs 3%), the immobilized microbe system has several advantages:

1. Newly attached cells produce mainly one product, butanediol, (4–5 mg/hr/ gr gel), and a small amount of acetone, (0.3 mg/ml), whereas free cells form larger amounts of butanediol (9 mg/ml) but almost six times more acetone (1.8 mg/ml).

2. After reactivation, immobilized microbes can keep synthesizing butanediol continuously and at lower substrate concentrations compared with free cells (2.5% vs 10%).

3. Product yield of the microbial catalyst remains rather constant (0.5–0.6 mole product/mole substrate) and only a little lower than for free cells (0.75 mole product/mole substrate).

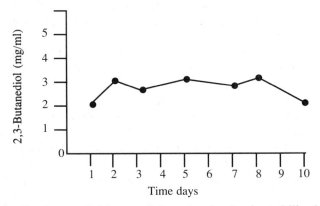

Figure 4 Continuous 2,3-butanediol production by immobilized *E. aerogenes* cells.[11]

4. It was shown that the *E. aerogenes* strain can produce butanediol in the presence of high salt concentrations. Since these salts increase the operational stability of carrageenan gel particles, successful substrate transformation can be performed on a very complex and rich medium. By properly matching carbon, nitrogen, and phosphate sources, one can expect higher 2,3-butanediol yield, increased cell concentration, and greater support stability.

Long term operational stability of the immobilized cell system, microbe reactivation in the entrapped state, acetone inhibition by the matrix, as well as lower sensitivity to pH, temperature, glucose and oxygen concentrations compared to free cells, strongly point to the feasibility of a continuous flow process for 2,3-butanediol synthesis using immobilized microbial cells.

6.2 N-butanol and Isopropanol Production by Living Immobilized *Clostridium butylicum* Cells[12]

N-butanol and isopropanol can be produced from oil and via fermentation, which can proceed as a batch or continuous operation using free cells or as a continuous process with immobilized microbes.

The efficiency of continuous free cell conversion was 20% higher than for batchwise transformation, but maximum duration of this activity was only 90 hours. On the other hand, during 215 hours of operation, productivity of the immobilized cell column was four times greater compared with free cells.

Clostridium butylicum LMD 27.6 cells were immobilized by entrapment into Ca-alginate gel, with an initial concentration inside the matrix of 9.4% (gr wet cells/gr gel). The beads (1mm diameter) were packed into a water jacketed glass column (37°C) through which a complete medium (glucose 6%, yeast extract 1%, $CaCl_2 \cdot 2H_2O$ 0.5%) was pumped at the rate of 8.6 ml/hr, giving a residence time of 3.5 hours.

At the glucose concentration of 57 ± 2.5 gr/l, butanol and isopropanol were the main products formed, although negligible amounts of ethanol, acetone, acetic and butyric acids were detected.

N-butanol yield was about 30% of theoretical and the substrate conversion rate was found to be less than 40%. The authors attributed this small product yield to low strain activity. The authors also observed that independent of the glucose level in the medium, final product concentration never exceeded 5 gr/l. It was noted that two parameters strongly affected reactor performance: product inhibition and pH; a decrease in pH due to acid byproducts lowered product yield.

Even with these limitations, applying living immobilized microbial cells for n-butanol-isopropanol production has the great advantage of a four fold higher productivity compared with batchwise fermentation during 215 hours of operation.

6.3 Acetone-butanol Production by Living Immobilized *Clostridium acetobutylicum* Cells[13]

Presently, acetone-butanol synthesis is a two phase batchwise process which takes 30–48 hours. The first, logarithmic phase of growth, is accompanied by formation of acetic and butyric acids, causing a drop in pH to occur. During the second phase, the solvents are produced while microbial growth ceases. The maximum total solvent yield is usually 30% v/v of the carbon utilized. Maximum solvent accumulation depends on the type of strain and is on the order of 19 gr/l, with a butanol content of 7–9 gr/l.

Clostridium acetobutylicum ATCC 824 spores and vegetable cells were immobilized by entrapment into sodium alginate gel (2 gr gel/100 gr cells). Spores were first heat reactivated 95°C) and allowed to form mycelium inside the matrix for 7 hours at 35°C. Bead-like gel particles (2mm diameter) were then used in batchwise and continuous operations. To improve mechanical strength of the support, glutaraldehyde treatment was necessary.

6.3.1 Batchwise Fermentation Using Immobilized Vegetable Cells and Spores

Spores and cells in the logarithmic growth phase were immobilized for substrate (glucose) conversion via batchwise fermentation. The patterns of product formation by entrapped cells and spores were notably similar to those of free cells: after 24 hours of incubation, acetic and butyric acids were no longer synthesized; following 48 hours of incubation, the amount of butyric acid declined, while acetone, butanol, and acetic acid were detected. 0.82 gr/l vs 0.24 gr/l, 3.71 gr/l vs 1.81 gr/l, and 1.45 gr/l vs 0.26 gr/l were the values for acetone, butanol, and acetic acid production by cells immobilized in the logarithmic growth phase compared with immobilized spores, respectively. Entrapped spores were chosen for further study because a dominant product was formed (butanol).

Adding butyric acid to the medium caused greater formation of butanol (2.89 gr/l), and negligible amounts of acetic acid (0.47 gr/l) and acetone (0.61 gr/l), resulting in a 4.7:1 butanol:acetone ratio. It appeared that the butyric acid was consumed first, and only after that was glucose assimilated.

6.3.2   Repeated Batchwise Operation Using Immobilized Spores

Repeated batchwise fermentation was performed by *C. acetobutylicum* spores immobilized via entrapment into sodium alginate gel. Three consecutive runs lasting 20–24 hours each were carried out using a substrate consisting of glucose (10 gr/l) and butyric acid (3 gr/l), (pH = 4.5). Although acetone and acetic acid were found in the solution, butanol was the main solvent synthesized, with a productivity of 2.32 gr/l − 3.81 gr/l. Product yield was 12.3–20.1%, and the butanol: acetone ratio in the mixture was 5:1.

6.3.3   Continuous Column Operation Using Immobilized Spores

Solvent production was also studied in a continuous column reactor (125 ml volume, 35°C) packed with *C. acetobutylicum* spores entrapped into the Ca-alginate gel. After spore growth inside the matrix (35°C, overnight), bead shaped particles (2mm diameter) containing vegetable mycelium were packed into the reactor. A complete medium from a reservoir was circulated through the column for 24–48 hours, which allowed entrapped cells to reach the solvent production phase. After 24–48 hr, mycelial growth inside the gel was interrupted by washing the beads. Substrate containing glucose (3%), butyric acid (0.3%), and salts (pH = 4.5) was then pumped through the column at a medium flow rate of 137 ml/hr. This system formed several products: butanol, acetone, ethanol, and acetic acid. During steady state, the acetone and butanol concentrations were 0.37 gr/l and 2.05 gr/l, respectively, giving a butanol: acetone ratio of 5.5:1. The amount of ethanol synthesized was less than 10% of butanol production; some acetic acid (0.7 gr/l) was also accumulated.

The author calculated that immobilized cell reactor activity peaked after 96 hr of operation, yielding 57 gr butanol/l of reactor volume per day. However, during long term continuous conversion, the rate of butanol formation dropped to 20– 25 gr/l of reactor volume per day, and was nearly stable at that level over 500–800 hr of operation (Figs. 5, 6).

Thus, the data show that by using entrapped *C. acetobutylicum* cells, butanol can be fermented on a continuous basis with a productivity several times higher than for free cells.

6.4   Summary

Applying immobilized microbial cells for solvent production (ethanol, 2,3-butanediol, n-butanol and isopropanol, acetone-butanol) presents the most recent alternative to conventional fermentation processes.

The major advantage of attached cells is that after entrapment into support

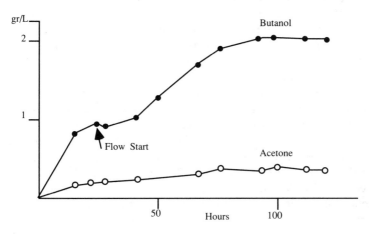

Figure 5   Short term continuous acetone-butanol production by immobilized *C. acetobutylicum* cells.[12]

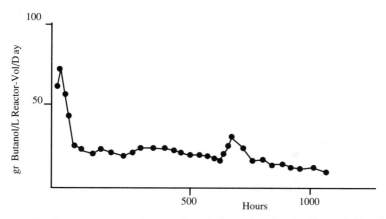

Figure 6   Long term continuous butanol production by immobilized *C. acetobutylicum* cells.[12]

matrices, they remain alive, able to multiply, carry out multistep enzymatic reactions, regenerate cofactors, and can be reactivated. Effectiveness of a microbial catalyst depends on properties of the cells and the carrier, the method of immobilization, availability of carbon and nitrogen sources, medium flow rate, etc. Each of the components contributes to catalyst performance, and when properly matched, the product yield as well as operational stability of the system can be higher than for free cells.

Various systems were tested for ethanol fermentation; immobilized yeasts and immobilized bacteria are the most attractive from the industrial standpoint since they provide greater productivity, operate at higher substrate concentrations and at shorter residence times. Also, immobilized bacteria perform better than immobilized yeasts at high medium flow rates.

Application of attached microbes for 2,3-butanediol synthesis is more efficient compared with using free cells because the butanediol yield is high, while acetone production is inhibited by the support matrix.

In the case of n-butanol-isopropanol, substrate transformation via immobilized cells increases the yield by a factor of four.

Acetone-butanol formation can be performed by attached cells or spores in repeated batch and continuous operations. In the latter case, when butyric acid and glucose are used as substrates, butanol is the major product synthesized. The acetone-butanol ratio in this instance is 5.5:1.

All of the systems described exhibit high operational and storage stabilities.

The results strongly suggest that application of immobilized living microbial cells is very advantageous compared to already existing technologies.

Literature

1. Cysenski, G., Wilke, C., *Biotech. Bioeng.*, **20**, 1421 (1978)
2. Chibata, I., US Patent 4,138,292 (1979); in *Immobilized Microbial Cells* K. Venkatasubramanian (Ed.), *ACS Symposium*, Series 106, 187 (1979)
3. Wada, M. *et al.*, *Eur. J. Appl. Micro. & Biotech.*, **10**, 275 (1980), **11**, 67 (1981)
4. Hagerdal, B., Mosbach, K., Paper presented at Int. Conf. Eng. and Food, Helsinki, Finland, 1979
5. Navarro, J. M., *et al.*, *Industries alimentaries et apricoles*, **6**, 695 (1976)
6. Rogers, P. *et al.*, *Proc. Biochem.*, **15**, 6, 7 (1980)
7. Lee, K. J. *et al.*, *Proc. Biochem.*, **2**, 11, 487 (1980)
8. Awari, E. J. *et al.*, *Biotech. Letters*, **2**, 11, 499 (1980)
9. Grote, W., *et al.*, *Biotech. Letters*, **2**, 11, 481 (1980)
10. Amin, G., *et al.*, *Eur. J. Applied Microb.*, **14**, 59 (1982)
11. Chua, J. M., *et al.*, *J. Ferm. Tech.*, **58**, 2, 123 (1980)
12. Krouwell, P. G., *et al.*, *Biotech. Letters*, **2**, 5, 253 (1980)
13. Häggstrom, E., Molin N., *Biotech. Letters*, **2**, 5, 241 (1980); *Adv. Biotech.*, **2**, 79 (1981)
14a. Awari, E. *et al.*, *Chem. Eng. News*, Oct 29, 12 (1979)
14b. Awari, E. *et al.*, *Chem. Eng. News*, Oct 8, 19 (1979)
15. Hahn-Hägerdale, B. *et al.*, *Appl. Biotech. Bioeng.*, **7**, 43 (1982)

16. Hahn-Hägerdale, B. *et al.*, *J. Chem. Tech. & Biotech.*, **32**, 15, 7 (1982)
17. Hahn-Hägerdale, B. *et al.*, *Biotech. Bioeng. Symp.*, **11**, 651 (1981)
18. Hahn-Hägerdale, B. *et al.*, *Biotech. Letters*, **3**, 2, 53 (1981)
19. Lipinsky, E. S., *Chem. Eng. News*, June 22, 54 (1981)
20. Lipinsky, E. S., *Science*, **212**, 1465 (1981)
21. Lipinsky, E. S., *Chem. Eng. News*, Nov. 3, 35 (1980)
22. Skotincki, M. L., *Appl. Environ. Micro.*, **40**, 1, 7 (1980)
23. Hahn-Hägerdale, B. *et al.*, *Desalination*, **35**, 365 (1980)
24. Erikson, K. E., *Biotech. Bioeng.*, **20**, 317 (1978)
25. Grethlein, H. E., *Biotech. Bioeng.*, **20**, 503 (1978)
26. Menezes, T. J., *et al.*, *Biotech. Bioeng.*, **20**, 555 (1978)
27. Okazaki, M., Moo-Young, M., *Biotech. Bioeng.*, **20**, 637, 1978
28. Ladish, M. R. *et al.*, *Science*, **201**, 743 (1978)
29. Tsao, T., *Proc. Biochem.*, **13**, 10, 12 (1978)
30. Humphrey, A. E., in *Biotechnology and Bioengineering Symposium*, Gaden, E. L., Humphrey, A. E. (Eds.), **7**, 45 (1977)
31. Wilke, C. R. (Ed.), *Biotechnology and Bioengineering Symposium*, **5** (1975)
32. Mozes, N. *et al.* *Appl. Micro. & Biotech.*, **22**, 2, 92 (1985)
33. Jain, V. *et al.*, *Biotech. Bioeng.*, **27**, 3, 273 (1985)
34. Keller, E., Lingerrs, F., *Appl. Micro. & Biotech.*, **20**, 1, 3 (1984)
35. Krysteva, M., Blagov, S., Sokolov, T., *J. Appl. Biochem.*, **6**, 5–6, 367 (1984)
36. Cantarella, M. *et al.*, *Appl. Micro. & Biotech.*, **20**, 4, 238 (1984)
37. Gibbons, W., *Biotech. Bioeng.*, **25**, 9, 2127 (1983)
38. Dekker, R., *Biotech. Bioeng.*, **25**, 4, 1127 (1983)

7

Microbial Catalysts for Steroid Transformations

The purpose of this chapter is to compare various methods for steroid modifications, emphasizing the most recent trend: application of immobilized living microbial cells for transformations of industrial importance: Δ'-dehydrogenation, 11α, 11β, 16α-hydroxylations, and side-chain cleavage at the C_{17}-position. Properties of different supports and their retention capacities are presented. The effects of solvent polarity and matrix contribution to catalyst performance are evaluated. Methods of immobilization and mechanisms of cell-support interactions are discussed. Steroid Δ'-dehydrogenation in organic solvents and in column type reactors is compared with catalyst performance in batchwise operation at high substrate concentrations. The effect of support matrices on 11α and 16α-hydroxylations is presented. 11β-hydroxylation by immobilized cells and immobilized spores at different substrate concentrations is compared. Side-chain cleavage at the C_{17}-position in the presence of inhibitors and on modified substrates in batchwise and continuous operations is evaluated. Factors limiting transformations in column type reactors are discussed. Changes in cell properties after immobilization (temperature-pH optima, operational stability, and reactivation in the immobilized state) are presented.

Introduction

The diminishing supply of steroid raw materials, increasing demand for anti-inflammatory agents, contraceptives, and sex hormones are resulting in the development of alternate sources of steroids as well as new methods for steroid modifications. These methods can be divided into total chemical synthesis and transformations of naturally occurring sterols already containing typical ring structures.

Total chemical synthesis is outside the scope of this chapter, and despite the progress made in the field, it represents a difficult process due to numerous steps and the great number of isomers formed. The problem is aggravated by devising stereospecific reactions in which a desired isomer predominates in the final product, and resolution of racemates. For example, the total chemical synthesis of cortisone acetate from 3-ethoxypenta-1,3-diene and benzoquinone involves several intermediates and numerous subsequent reactions.[1,2]

Transformation of naturally occurring steroids such as diosgenin and plant sterols like stigmasterol or cholesterol to pharmaceuticals can be accomplished by enzymatic as well as microbiological methods and performed by free enzymes (cells) or immobilized enzymes (cells). In the latter case, the enzymes (cells) can be attached to numerous supports by various techniques.

Depending on the type of transformation, the process may be governed by a single enzyme or by multiple enzymes. These catalysts may be inducible or constitutive.

Certain steroid transformations require an enzyme and a cofactor, which must be added to the system. Due to their expense, two different mechanisms for cofactor regeneration were proposed.[3] The first consisted of co-immobilizing three enzymes (steroid dehydrogenase, alcohol dehydrogenase, and aldehyde dehydrogenase) to the support via glutaraldehyde. Steroid transformation was accomplished in the presence of NAD and ethanol. The alternative was to use immobilized whole cells, which could regenerate the cofactor, allowing conversion to proceed on a continuous basis at a much cheaper and more efficient rate.

The purpose of this chapter is to compare different methods of steroid transformation, their advantages and limitations, and to emphasize the most recent development in the field: application of microbial catalysts to reactions of industrial importance (Δ'-dehydrogenation, 11α, 16α, 11β-hydroxylation, side-chain cleavage at the C_{17}-position). The review will be based on most recent publications and patents from USA, Sweden, Japan, Germany, etc.

7.0   Transformations by Free or Immobilized Enzymes

Using free or immobilized enzymes for steroid transformations represents an early approach.[4] *Arthrobacter simplex* cells were grown on a suitable medium, and the Δ'-dehydrogenase enzyme was induced by adding cortisol. Cells were collected and lysed; the supernatant formed exhibited 400–600 mU/ml of enzyme activity. Production was increased ten fold by eliminating nucleic acids and subsequent ammonium sulfate precipitation of the supernatant.

Free Δ'-dehydrogenase was completely inhibited by mercuric chloride, potassium ferrocyanate, copper sulfate, and reagents with thiol or disulfite groups. The enzyme was activated in the presence of methanol (3–6%) or a smaller amount of ethanol (1–2%).

An immobilized preparation was formed via entrapment of the free enzyme into acrylamide gel, followed by photopolymerization. This method of attachment was chosen in order to minimize denaturation.[5]

Both, free and immobilized enzyme preparations catalyzed conversion of cortisol to prednisolone only in the presence of external cofactors such as menadione, phenazine methosulfate, or various quinones. Furthermore, the catalyst retained only seven percent of the initial activity of free enzymes, and the rate of prednisolone formation fell to only 9mM/hr/gr of polymer. Despite the low productivity, immobilized enzymes were stable at a low temperature (+4°C) and maintained their activity for a four week period.

It was shown[6] that immobilized enzyme preparations formed from equal amounts of cells as immobilized microbe catalysts, exhibited a ten fold lower enzymatic conversion than their counterparts.

Steroid transformation by free and attached enzymes had several disadvantages:
1. Enzyme isolation and purification usually resulted in loss of activity.
2. Reaction could not proceed without an expensive electron acceptor.
3. A special system was necessary for cofactor regeneration.
4. Immobilized enzymes did not maintain sufficient production, which was even lower than that of unattached enzymes.

Since the above limitations could not be overcome, the attractiveness of the enzymatic process for steroid transformation was very limited.

7.1   Microbial Transformations

Microbial transformations can be performed by free and resting cells (spores) in batch (repeated batch) operations; by immobilized cells (spores) in continuous fermentation in the presence of organic solvents; or via repeated batch operations at high substrate concentrations (pseudo-crystallofermentation).

7.1.1   Transformation by Free Cells[8,9]

The majority of steroid transformations are still carried out as batch processes. Enzyme induction is usually accomplished by adding steroid to the microbes and growing the cells until the highest rate of activity is reached.

Substrate dissolved in solvents (ethanol, methanol, butanol, dimethylforma-mide, propylene glycol, etc.) is then added to the culture and incubated with the mycelium for several days. The product formed is extracted and purified. Three major problems are encountered in the free cell system: low substrate solubility, solvent toxicity, periodicity of the processes. To enhance substrate solubility, various concentrations of different solvents were tried. Some, however, were toxic to the cells, indicating that the type of solvent, time and temperature of exposure to the culture should be selected very carefully. Important microbial transformations that are still industrially performed by free cells in batchwise operations are presented in Table 1.

Since the periodicity of batchwise conversion by free microbes can not be overcome, and due to high operational stability of immobilized cells under the same conditions, the inevitable trend is toward applying attached cells to batchwise processes performed at high substrate concentrations and to con-tinuous flow operations in the presence of organic solvents.

7.1.2   Transformation by Resting Cells[10]

This method of steroid transformation has two advantages over free cell fermentation: a faster conversion time and greater system stability.

A microbial culture is first grown on a steroid free medium. The cells are then resuspended in water (or buffer), to which the substrate is added. This way no enzyme induction is involved, and the synthesis results from activity of enzymes formed during the earlier growth stage without exogenous steroids.

Resting cells of *Sepedonium ampullosporum* mold were used for hydroxyla-tion of 9β,10α-pregna-4,6-dien-3,20 diol to 6-dehydro-16α-hydroxy-9β,10α-progesterol. Microbes were suspended in distilled water, phosphate, or tris buffer. After twelve hours of incubation, steroid conversion reached 76%, compared to 46% for free cells in batch operation. Complete substrate fermentation was achieved in about 50–70% of the time needed for freely suspended cells. A third advantage of the resting cell system is that it can be semicontinuously exposed to steroid. Up to twelve sequential doses of substrate, each 0.25 gr/liter, were converted to product, yielding 71% of theoretical productivity. Factors affecting the rate of fermentation were pH (optimal 6.0–7.0), temperature (optimal 25–30°C), aeration, and agitation. The hydroxylation system was found to have a critical oxygen concentration above 90% of air saturation in water. The operation was stable for 150 hours.

Table 1 Transformations by Microbial Catalysts

Substrate	Product	Single Enzyme with Cofactor/Multiple Enzyme	Microbe Used	Method of Immobilization	References
Cortisol	Prednisolone	Δ'-dehydrogenase/	*Corynebacterium simplex* ATCC 6946	Entrapment into acrylamide gel	11, 12, 13
Hydrocortisone	Prednisolone	Δ'-dehydrogenase/	*Corynebacterium simplex*	Entrapment into collagen	14, 15
Hydrocortisone	Prednisolone	Δ'-dehydrogenase	*Mycobacterium globiforme* 193	Entrapment into acrylamide gel	16
20-oxy-hydrocortisone	Hydrocortisone	20β-dehydrogenase/	*Mycobacterium globiforme* 193	Entrapment into acrylamide gel	17
Reichstein substance S (RCS)	Δ'-RCS	Δ'-dehydrogenase/	*Pseudomonas testosteroni* ATCC 11996	Entrapment into acrylamide gel	18
Reichstein substance S RCS	Cortisol	11β-hydroxylase	*Curvularia lunata*	Entrapment into acrylamide gel	5
Cholesterol derivatives	Androstenedione derivatives	/Side-chain cleavage	*Mycobacterium phlei* IMET SG 1026	Entrapment into acrylamide, Al-alginate, Ca-alginate	19a, b, c
Cholesterol derivatives	Androstenedione derivatives	/Side-chain cleavage	*M. smegmatis* IMET H-124	DEAE, alginate, acrylamide	20
Testosterone	4 AD, DTS, ADD	/Multienzyme	*Nocardia rhodocrous* NCIB 10554	Entrapment into H or L gels	22

7.1.3 Transformation by Immobilized Cells

Applying immobilized cells for steroid transformations has several advantages over immobilized enzymes, free cells, and resting cells because

1. The need to isolate and purify enzymes is eliminated.
2. There is a lower or no requirement for cofactor addition.
3. Enzymes are more stable inside the cells.
4. The operational stability of immobilized microbes is longer.
5. By contrast with free cells, microbial matrices are reusable and suitable for continuous operation.
6. If the support and method of immobilization are properly chosen, a catalyst can perform various steroid transformations in organic solvents, in the presence of inhibitors, and on modified substrates.
7. Attached cells can be reactivated inside the matrix.

Table 1 lists various steroid transformations carried out by immobilized cells on a laboratory scale.

The following factors affect cell performance:

1. The concentrations of inducers (if enzymes are inducible).
2. Support properties and retention capacities.
3. The method of cell immobilization.
4. Substrate/product concentrations.
5. Properties of the solvent used.
6. Types and concentrations of inhibitors.
7. The oxygen concentration (which dictates the type of reactor).

7.1.4 Supports for Immobilized Cells

Numerous supports were used for attachment of cells performing steroid transformations, including polyacrylamide gel, Al- or Ca-alginate gels, Fe^{+3} hydroxigel, photocrosslinkable resins, urethane prepolymers, collagen membranes, various types of cellulose, inorganic supports such as aluminum oxide, and activated silica. The methods used for immobilization were primarily entrapment and adsorption. The mechanisms responsible for cell attachment varied from free radical polymerization (acrylamide) to electrostatic interactions (amberlite, silica), weak ionic bond formation (collagen), and simple entrapment (alginate gel, various membranes) (Table 2).

The retention capacity is one of the most important characteristics of a support. Gels with high microbial counts are preferred because they exhibit greater catalytic activities.[6,20,26] The effect of cell concentration on the transformation capacities of matrices is presented in Tables 3, 4.

Table 2 Supports for Steroid Transformations

Type of Support	Method of Immobilization	Mechanism	Retention Capacity	References
Acrylamide gel (2–15%)	Entrapment	Free radical polymerization	50%	12, 24
Hydrophobic (H) gels PU-6	Entrapment			21
Lipophilic (L) gels PU-3, ENTR-2000	Entrapment			21
Urethane Prepolymers (PU 1–10)	Entrapment			23
Ca-alginate gel	Entrapment		80–90%	24
Collagen	Entrapment	Ionic & electrostatic interaction	50%	14
Al-alginate gel	Entrapment			19b, 24
Fe^{+++} hydroxigel	Entrapment			19b, 24
Fe^{+++} hydroxigel/ gelatine	Entrapment			19b, 24
Sepharose activated by CNBr	Adsorption	Electrostatic interactions		19b, 24
Cellulose (powder)	Adsorption	Electrostatic interactions		19b, 24
Cellulose (crystals)	Adsorption	Electrostatic interactions		19b, 24
DEAE cellulose	Adsorption	Electrostatic interactions		19b, 24
Glasswool	Adsorption	Electrostatic interactions		19b, 24
Amberlite	Adsorption	Electrostatic interactions		19b, 24
Kieselgel	Adsorption	Electrostatic interactions		19b, 24
Al_2O_3	Adsorption	Electrostatic interactions		19b, 24
Activated silica gel	Adsorption	Electrostatic interactions	30–40 mg/gr	25

Table 3   Effect of Cell Concentration Inside the Support on Δ'-dehydrogenase Activity[26]*

Amount of Biomass Inside the Support (mg/ml)	Transformation Capacity of the Catalyst (%)		
	Hydrocortisone	Prednisolone	20β-oxy Derivative of Prednisolone
0.5	100	Traces	—
1.0	95	5	—
2.0	80	10	—
4.0	50	50	—
8.0	4	95	0.5
16.0	2	96	2.0
1.0 (free cells)	—	100	—
16.0 (free cells)	—	65	5.0

*Microorganism: *Mycobacterium phlei 193*; Support: Membranes; Method of immobilization: Entrapment

Table 4   Effect of Cell Concentration Inside the Support on Side-Chain Cleavage at C_{17}-position[20]*

Amount of Biomass Inside the Support (gr cell/gr support)	Transformation Capacity of the Catalyst (%)	
	First Batch	Second Batch
0.5	5	10
1.0	30	65
2.0	40	80
3.0	96	95

*Microorganism: *Mycobacterium phlei* IMET SG 1026; Support: Acrylamide gel

A rise in cell density inside the support from 1 mg/ml to 16 mg/ml caused an increase of 91% in Δ'-dehydrogenase activity. Up to 96% prednisolone yield was achieved at cell concentrations ranging from 8 to 16 mg/ml. With high microbe counts, a small amount (0.5–2%) of β-oxy derivative was formed in addition to the product (Table 3).

The data presented in Table 4 illustrate the effect of cell mass inside the support on side-chain cleavage of sterols at the C_{17}-position. Although the absolute numbers are different than for the Δ'-dehydrogenase enzyme, the relationship is similar: At a low biomass concentration (0.5 gr cells/gr support), side chain cleavage of the catalyst is only 5%, but an increase in cell density to 3 gr/gr support raises the product yield to 96%.

Due to the highly hydrophobic characters of substrates and products in steroid transformations, special attention should be paid to the nature of the support. It was suggested that cell entrapment into highly lipophilic carriers may increase substrate diffusion through the gel matrix, allowing entrapped microbes to form more effective catalysts.[22] *Nocardia rhodocrous* NCIB 10554 cells were immobilized into lipophilic (L) and hydrophilic (H) gels, and the complexes formed were used for testosterone (TS) transformation. It was found that depending on the type of support, two different routes for product formation (ADD: 1,4-androstadiene-3,17-dione) could be selected. Cells entrapped into the lipophilic (L) gel synthesized 4-androstene-3,17-dione (4-AD) during the following pathway to the final product: TS→4-AD→ADD (the last step is dehydrogenation). Microbes entrapped into hydrophilic (H) gels mainly produced dehydrotestosterone (DTS) via the second path of product formation: TS→DTS→ADD. In both methods, only Δ'-dehydrogenation of TS and 4-AD required phenasine methosulfate (PMS), whereas 17β-dehydrogenation of TS and DTS occurred without it. The reactions with L-gels proceeded via route one because only these matrices exhibited affinity to PMS. The authors concluded that selection of the method of conversion is governed by different affinities of PMS to various supports. Thus, selective formation of a desired product can be influenced by the hydrophilic or lipophilic character of the carrier.

Arthrobacter simplex cells were entrapped into eleven urethane prepolymers differing in molecular weight and ethylene oxide content.[23] A mixture of the three polymers which exhibited highest steroid transformation (PU-3, PU-6, and PU-9) was prepared, giving gels with various hydrophobicities. This allowed tailoring supports with specific chemical and physical properties.

When choosing a matrix for a catalyst, the specific properties of the cell and support should be carefully considered. For example, if acrylamide gel is used, one should remember that this carrier is toxic to cells, and therefore polymerization should proceed as quickly as possible at a low temperature. If calcium alginate gel is chosen, a concentration higher than 4% cannot be used due to difficulties in handling a material with high viscosity. Some urethane prepolymers (PU-5) do not form strong gels, whereas others (PU-10) crosslink too rapidly. On the other hand, titanium or chromium activated inorganic supports like silica do not form stable immobilized cell complexes. Their half lives are only six days, and their activities fall drastically during repeated operation due to cell desorption from the support surfaces.

The above data further emphasize the importance of matrix contribution to catalyst performance.[33a–d]

7.2 Δ'-dehydrogenation

Incorporation of a double bond into the 1–2 position of steroid molecules by immobilized microbial cells was studied in continuous flow reactors with substrate dissolved in organic solvents, and/or using pseudo-crystallofermentation (batchwise operation at high substrate concentrations). Since batch fermentation could proceed at high substrate/product concentrations, the inducible Δ'-dehydrogenase enzyme differs from many others by not being subject to substrate inhibition. The most recent research work shows a rather complicated substrate-product relationship: At low substrate concentrations, the product inhibits enzyme formation; at high substrate levels, the enzyme is first stimulated and then inhibited by increasing levels of product. This behavior is explained by the formation of a tertiary intermediate during the reaction.[15]

Prior to immobilization, Δ'-dehydrogenase activity in the cells is induced by adding substrate dissolved in organic solvents. Activated microbes are then attached by different techniques. Investigators in the USA, Sweden, Japan, as well as other countries are studying steroid Δ'-dehydrogenation by catalysts on a laboratory scale. Each group uses the same microbe-dehydrogenator, *Arthrobacter simplex*, but utilizes different supports and methods of immobilization. In this chapter, the results of these researches will be compared and discussed, emphasizing 1. methods of attachment, 2. matrix contributions, 3. effect of the solvents, 4. types of reactors, 5. yields, operational stabilities of the systems, and 6. cell activation in the immobilized state (Table 5).

The importance of support selection and the choice for the method of immobilization were discussed in detail earlier (Chapters 1–4).[33a–d] Microbe and carrier properties should be matched properly, as this affects the operational stability of the system. From Table 5, one can see that the entrapment technique is more advantageous compared to adsorption because the half-life of the carrier can be significantly longer (140 days for cells entrapped into acrylamide gel vs 6–12 days for cells adsorbed to cellulose or activated silica gel). The shorter half-life of the catalyst formed by adsorption is explained by cell desorption from the support.[25] Entrapment also usually results in higher support retention capacities. Despite acrylamide monomer toxicity to the cells, if the following precautions are taken, immobilization will not damage the microbes[7,12,13]:

1. The time between mixing the monomer with cells and the start of polymerization is minimized (1 min).
2. The polymerization process is performed at a low temperature (0°C).
3. The polymerization medium is buffered at pH = 7.5.

If polymerization is performed in an unbuffered solution, or the time of contact with the monomer increases, or even if the temperature rises, the viability of cells inside the support and microbial activity significantly diminish. In

Table 5 Effect of Different Modes of Operation on Δ'-steroid-dehydrogenation*

Support	Method of Immobilization	Reactor Type	Solvent	Half-life (days)	Temperature (°C)	Product Yield (%)	References
Acrylamide	Column	Entrapment	Methanol		30		13
		Batchwise†	—	40	30	100	13
Collagen	Entrapment	Column	Ethanol	5.43	30	85	14
		Batchwise†	—	4.0	30	80	15
Polyvinyl chloride membrane	Entrapment	Column	Methanol	7.0	30		26
Acrylamide	Entrapment	Batchwise†	—	140.0	30	60–70	17
Cellulose	Adsorption	Batchwise	—	12	30		25
Activated silica gel	Adsorption	Batchwise	—	6	30		25
Urethane prepolymer PU-3; PU-9	Entrapment	Batchwise	—	30	30	0.230	23
				operations		μM/h/gr	

*Microorganism: Arthrobacter simplex
†Batchwise fermentation proceeded with high substrate concentrations. Cells were reactivated in the immobilized state.

addition to high retention capacities, cell entrapment into any of the supports (collagen, calcium alginate, urethane prepolymer) does not result in cell leakage from the matrix.

7.2.1   Effects of Solvents

The most significant difference between steroid transformations and other processes is that the substrates and products are water insoluble. To increase substrate solubility and product yield, various solvents were applied, but this added a variable to the system by necessitating consideration of the effect of solvent on cells as well as on the support matrix. Thus, the transformation capacity of a steroid catalyst depends not only on the enzymatic activity of a microbe, the method of cell immobilization, the retention capacity of the support, and nature of the matrix, but also on the type of solvent used and the duration of cell exposure.

The effects of various organic solvents on Δ'-dehydrogenase activity are presented in Table 6. At a concentration from 5 to 10% and a treatment time of thirty minutes, methanol increased the activity of immobilized A. simplex cells by 10–20% using acrylamide, and by 32% using urethane prepolymers as supports. Raising the alcohol concentration to 15% caused a drop in microbial activity, which may have resulted from solvent toxicity to the cells.

Low quantities of butanol (0.1–0.5%) increased the transformation capacity by 25–30% when acrylamide was used as the support, but higher butanol levels (10%) caused a complete loss of catalytic activity.

Ethanol concentrations from 5 to 20% also increased product synthesis when collagen was used as the matrix. However, only a 10% concentration of the same solvent lowered synthesis by the urethane prepolymer PU-9 matrix.

Toluene is toxic to cells even at low quantities (0.1–0.5%), and at lengths of exposure as short as 5 minutes. The most recent results on the effects of organic solvents are encouraging. A. simplex cells were immobilized into the urethane prepolymer PU-9, and ethylene glycol, propylene glycol, or glycerol were used as solvents. Activity of the immobilized cell catalyst was increased by 40–90%.

The data above suggest a direct interaction between the solvent, cells, and support, and point to a possible protective effect of some matrices. Organic solvents were applied only in developing continuous flow processes for steroid transformations. The results presented show the feasibility of increasing system effectiveness by properly matching cell, support, and solvent properties.

Table 6 Effects of Organic Solvents on Steroid Dehydrogenase of Immobilized A. simplex Cells*

Solvent	Concentration (%)	Support	Length of Solvent Treatment (min)	Activity	Conversion Rate (%)	References
Methanol	5	Acrylamide	30	120		13
	10	Acrylamide	30	110		13
	15	Acrylamide	30	70		13
Butanol	0.1	Acrylamide	10	125		13
	0.1	Acrylamide	20	133		13
	0.1	Acrylamide	30	130		13
	0.5	Acrylamide	10	130		13
	10.0	Acrylamide	10	0		13
Toluene	0.1	Acrylamide	5	20		13
	0.5	Acrylamide	5	30		13
Ethanol	5.0	Collagen			23	14
	10.0	Collagen			20	14
	15.0	Collagen			35	14
	20.0	Collagen			20	14
	30.0	Collagen			15	14
Methanol	10.0	Urethane prepolymer PU-9		132		23
Ethanol	10.0	Urethane prepolymer PU-9		85		23
Ethylene glycol	10–30	Urethane prepolymer PU-9		140		23
Propylene glycol	10.0	Urethane prepolymer PU-9		165		23
Trimethylene glycol	10.0	Urethane prepolymer PU-9		124		23
Glycerol	10.0	Urethane prepolymer PU-9		191		23

*Δ′-dehydrogenase enzyme activity of non-treated immobilized cells is set to 100%.

7.2.2   Effects of Solvent Polarity and Matrix Hydrophobicity on Catalyst Performance

It was mentioned that steroid transformations can proceed only in various organic solvents which dissolve the substrate and make it available to the immobilized cells. The solvents should also cause minimal denaturation of enzyme(s) responsible for substrate transformations.

As recent research suggests, more active microbial catalysts can be formed by properly matching solvents with various degrees of polarity to supports with different hydrophobicities.

Depending on the type of substrate, free *Nocardia rhodocrous* NCIB10554 cells performed the following steroid dehydrogenations: I. cholesterol → cholestenone, II. prednenolone→progresterone, III. dehydroepi-androsterone →4-androstene-3,17-dione.

Transformation I was performed by free cells in a mixture (1:1) of n-heptene with nine solvents of different polarities (non-polar, moderately polar, and extremely polar). It appeared that substrate conversion declined with increasing solvent polarity. The highest transformation capacities (57–68%) were determined in the non-polar solvents, and no activity was found in extremely polar solvents.

Immobilized cell preparations were formed by *N. rhodocrous* entrapment into a hydrophilic resin (ENT-4000). The catalysts were placed into the same nine solvents as free cells, and steroid transformation was examined.

Of the nine solvents tested, substrate fermentation was observed in only two cases: chloroform-n-heptene (1:1, moderately polar), and methylene chloriden-heptene (1:1, polar). Attached cells were not active using all non-polar solvents, in which free cells showed highest productivity. In another experiment, immobilized microbe preparations were made by cell entrapment into ten supports of different hydrophilic and lipophilic natures. The transformation capacity of each catalyst was tested on substrates I, II, III, using two solvents with different polarities: A and B. (Solvent A was a mixture of benzenen-heptene, 1:1 v/v, non-polar; solvent B was a mixture of chloroform-n-heptene, 1:1 v/v, moderately polar.) In both solvents, prednenolone, a more hydrophilic steroid containing an acetyl group at the C_{17}-position, was more effectively dehydrogenated by cells entrapped in hydrophilic gels such as ENT-4000 and PU-6 than by the hydrophobic gel-entrapped microbes.

During transformation of cholesterol, a highly lipophilic steroid with an aliphatic side chain at the C_{17}-position, cells entrapped into lipophilic gels such as ENT P-2000, ENT B-1000, and PU-3 exhibited higher activity than cells entrapped in hydrophilic matrices.

The effect of the degree of solvent hydrophobicity was more drastic when solvent A, less polar than solvent B, was used for cholesterol conversion. Cells entrapped into hydrophilic prepolymers such as ENT-4000, PU-6, or a mixture of PU-6 and PU-3 (75:25; 50:50), could not catalyze substrate dehydrogenation in solvent A, but could perform the same reaction in solvent B.

Thus, it can be concluded that lipophilic substrates are more difficult to diffuse into the matrix of a hydrophilic gel using a non-polar solvent. Increasing solvent polarity reduces the diffusion barrier, permitting higher production by entrapped cells. However, extremely high polarity may drastically lower microbial activity. Also, the more lipophilic substrates are easier to diffuse into lipophilic gels by polar solvents.

The results indicate that solvent polarity affects the permeability of a cell matrix to a steroid. By properly matching this parameter with gel hydrophobicity and substrate polarity, the transformation capacity of a microbial catalyst can be comparable (or higher) than for free cells.

7.2.3   Column Operation for Steroid Δ'-dehydrogenation

Application of column type reactors for steroid transformations using immobilized microbial cells were investigated in Sweden, USA, and Japan (Table 5).

A group from the University of Lund (Sweden) used *A. simplex* microbes immobilized into acrylamide gel,[5,6,11–13] whereas researchers from Rutgers University (USA) used collagen entrapped cells.[14,15]

Collagen membrane chips (3 × 5mm) or acrylamide gel fractions (average size 0.2 mm) were packed into small water jacketed glass columns (capacity 2.5–6.0 grams of support). Substrate dissolved in ethanol (methanol) was pumped through the columns at various flow rates. For the collagen packed reactor, the optimal ethanol concentration was found to be 15%, and the optimal substrate concentration was 0.67 mg/ml. At a residence time of 30 min, the initial conversion rate was 79%, but dropped to 45% in 130 hours. The half-life of the system was 5.43 days.[14]

Researchers from Sweden examined the effect of oxygen on steroid transformation using 1mM substrate dissolved in 5% (v/v) methanol. Oxygen delivery to the column was accomplished by an in-line hollow-fiber gas permeator. At the concentration of 0.25 mM O_2, no change in steroid conversion was observed with various substrate flow rates, which suggested a necessity for extra oxygen. However, a high concentration inhibited product formation: dehydrogenase activity was lower in an oxygen vs air saturated medium. Nonetheless, at the substrate flow rate of 15 ml/hr and a higher oxygen level, the conversion rate reached 45–70%. By contrast with repeated batch fermentation via immobilized cells (discussed later), 100% substrate to product conversion was never reached during column runs. Two major disadvantages of this reactor were observed: The transformation capacity was five times lower, and the operational stability of the biocatalyst declined rapidly.

In view of such facts, and due to difficulties in supplying oxygen for column operation, repeated batchwise transformation at high substrate concentrations was a better process, as will be shown. However, most recent results indicate some future potential for a continuous flow system.[15]

7.2.4   Pseudocrystallo-fermentation for Steroid Δ'-dehydrogenation[12,13,15,27]

An alternative to continuous column transformation that avoids oxygen limitations and the effect of solvents is batchwise operation at high substrate concentrations, which was feasible only because there was no substrate/product inhibition. A microbial catalyst was first formed by cell entrapment into acrylamide or collagen. A quantity of steroid powder by far exceeding its solubility was next added to the catalyst medium. Matrix activity and the time necessary to convert 100% of the substrate are presented in Table 7.

With repeated runs, transformation capacity of the preparation increased ten fold, while the time for 100% substrate conversion declined at almost the same rate.

Table 7   Steroid Δ'-dehydrogenation by Immobilized Cells at High Substrate Concentrations[12]*

Repeated Runs	Transformation Capacity of the Catalyst (mg. prednisolone/hr/gr. wet gel)	Time for 100% Conversion
1	3.1	17.4
2	5.0	10.8
3	13.5	4.0
4	18.1	3.0
5	26.3	2.1
6	27.1	2.0
7	27.1	2.0
8	29.6	1.8
9	30.8	1.8
10	31.7	1.7

*Microbe: *Arthrobacter simplex*; Support: Acrylamide gel; Method of Immobilization: Entrapment. After each run, when all the substrate was converted, the gel was filtered off, washed, and incubated in freshly prepared nutrient medium, thus reactivating the cells inside the gel. (Repeated batch-wise operation.)

Due to high catalyst productivity, rapid substrate fermentation, and cell reactivation in the immobilized state, batchwise operation at high substrate concentrations is more attractive than the continuous flow process for steroid Δ'-dehydrogenation.

7.2.5   Cell Reactivation in the Immobilized State

Δ'-steroid dehydrogenase activity can be increased by cell reincubation in a nutrient medium with proper composition.[12,13,24]

From the previous discussion, it is clear that greater product yield can be obtained when steroid dehydrogenation is performed batchwise at high substrate concentrations than via a continuous flow process. The one disadvantage of crystallo-fermentation is a decline in the transformation capacity of the catalyst after repeated fermentations. This problem can be overcome by reincubating the matrix in a medium containing the proper combination of metallic ions and carbon/nitrogen sources.

The reactivating effect of these compounds on Δ'-dehydrogenase activity of immobilized microbial cells is presented in Table 8.

Catalyst activity declined upon incubation in water (used as control) or phosphate buffer (pH = 7.0) containing a mixture of $ZnCl_2/FeCl_2$ or $CoCl_2/MgSO_4$. Phosphate buffer without any metallic ions allowed retention of 60% of the initial transformation capacity after 16 days of operation. Reincubation in Tris-HCl buffer also resulted in retention of 70% of the transformation capacity after 10 days. The results thus suggest the importance of a buffer background for Δ'-dehydrogenase enzyme reactivation in the matrix. Incorporating traces of $CoCl_2$ or $MgSO_4$ into the phosphate buffer caused a complete loss of catalytic activity, whereas adding traces of $MgCl_2$ and $CaCl_2$ allowed 90% of the productivity to remain after 16 days of operation. Obviously, the presence of a buffer and certain metallic ions reactivated the enzyme inside the cells.

The effect of reactivation with nutrients is different from reactivation by buffer and salts because nutrients promote cell multiplication inside the catalyst. A glucose concentration of 0.2% raised the transformation two fold after 2–6 days of operation, and 90% of catalyst activity remained after 16 days. A low peptone concentration (0.01%) increased the rate of production, but did not have a prolonged effect, which was probably due to rapid utilization by cells. Higher peptone levels (0.1–1.0%) resulted in three to five fold greater transformation capacities and had prolonged effects (up to 16 days). A combination of peptone (0.5%), glucose (0.2%), phosphate buffer (pH = 7.0) and salts ($MgCl_2$, $CaCl_2$) raised catalyst activity by five to six fold and had a prolonged effect (up to 16 days). This suggests the importance of combining nutrients, buffer, and salts for reactivating immobilized microbial cells.

Table 8   Activating Effect of Buffers, Salts and Nutrients on Δ'-dehydrogenase Activity[2,3]

Medium (%)[†]	Initial Transformation Rate (%) After (days)*				
	0	2	6	10	16
Peptone					
0.01	100	125	30	0	0
0.1	100	200	225	165	120
0.5	100	460	500	650	530
1.0	100	290	320	380	550
Glucose (0.2)	100	210	170	110	90
Glucose (0.2), Peptone (0.01)	100	110	55	0	0
K_2HPO_4 (0.1M), pH = 7.0	100	70	50	60	60
Tris-HCl, 0.05M pH = 7.0	100	100	70	70	30
K_2HPO_4, 0.1M, pH = 7.0, $ZnCl_2$, $FeCl_2$	100	50	25	15	10
K_2HPO_4, 0.1M, pH = 7.0, $CoCl_2$, $MgSO_4$	100	40	40	0	0
K_2HPO_4, 0.1M, pH = 7.0, $MgCl_2$, $CaCl_2$	100	160	130	80	90
K_2HPO_4, 0.1M, pH = 7.0 Peptone—0.5, Glucose—0.2, $MgCl_2$, $CaCl_2$	100	560	650	600	550
H_2O	100	60	40	40	20

*The activity of freshly prepared gel is set to 100%
†The concentration of inorganic salts $MgCl_2$, $ZnCl_2$, $CoCl_2$, $FeCl_2$, $CaCl_2$, $MgSO_4$ was 1mM

The mechanism of cell reactivation in acrylamide gel was investigated using various methods, including application of chloramphenicol (an inhibitor of protein synthesis), benzylpenicillin (an inhibitor of cell wall synthesis), microscopic examination of stained cells, and gravimetric determination of gel weight. The authors rationalized that reactivation was a result of microbial growth inside the acrylamide matrix based on the facts that both antibiotics completely inhibited reactivation and that the immobilized cells could not be reactivated without nutrients. This conclusion suggests that if the necessary precautions are taken, cells can remain alive even after immobilization into toxic matrices. A recent publication from the same laboratory described how a microbial catalyst formed by cell entrapment into calcium alginate gel could also be reactivated in the immobilized state.[24]

7.3   11α-hydroxylation[34]

Incorporating a hydroxyl group into the 11α-position of a progesterone molecule is of major industrial importance since it represents a route to cortisone and other corticosteroids widely used in medicine. Although the reaction can be performed by free cells of several microorganisms: *Aspergillus ochraceus* NRRL 405, *A. niger* 12v, *Rhizopus nigricans* ATCC 6227b, and *R. nigricans* NRRL 1477, only the *R. nigricans* ATCC 6227b strain was immobilized.

Progesterone 11α-hydroxylase of *R. nigricans* ATCC 6227b is an inducible enzyme; thus cells were exposed to the substrate three hours prior to immobilization. Because concentration of the oxygen utilized in hydroxylation was critical, transformation by the catalyst was performed batchwise with shaking (150 RPM) in a buffer-salt solution (30°C) containing substrate dissolved in ethanol. In addition, 11α-hydroxylation by *R. nigricans* cells requires the cofactor NADPH, which was formed during metabolism of glucose (1%) especially added to the buffer-salt solution.

R. nigricans ATCC 6227b cells were immobilized by entrapment into three different gels: agar, alginate, and acrylamide. The type of support was found to significantly influence the transformation capacity of the catalyst: Highest activity (40–45%) was observed for cells incorporated into agar; alginate entrapped microbes exhibited a 2–3 fold lower product synthesis (12–19%); no activity was found in the catalyst formed with acrylamide, possibly due to toxicity of the monomer.

Substrate conversion by the immobilized microbial matrix depended on salts and nutrients in the medium, but not the buffer background. Performing fermentation in a buffer-salt solution yielded only 4–5mg of product, but adding 1% glucose raised the enzymatic activity by ten fold. Using lactalbumin as a nitrogen source (0.2%) caused a response similar to that of including glucose into the feed; substituting phosphate buffer for Tris buffer made no difference.

Fifty percent of substrate conversion was achieved in 24 hours. Unfortunately, catalyst activity declined after this period. To avoid inactivation and maintain 11α-hydroxylation in repeated operations, cell reactivation inside the support was carried out by placing twenty four hour old catalyst into a freshly prepared buffer-salt medium with 1% glucose and 2 mg/ml progesterone.

11α-hydroxylation by the catalyst was maintained for two repeated runs when the medium used for cell reactivation contained 1% glucose. If the reactivating medium included 1% glucose and 0.2% lactalbumin, catalyst activity could be maintained for four repeated runs. Increasing the lactalbumin concentration to 0.4% resulted in markedly lower enzyme formation. The data above point to the importance of glucose for NADPH regeneration and also to the glucose/nitrogen ratio for enzyme synthesis.

The cell concentration inside the support is known to significantly influence catalyst performance. For several steroid transformations (Δ'-dehydrogenation, side-chain cleavage at the C_{17}-position, etc.), a high cell concentration per unit support is desirable. Thus, applying the entrapment technique was preferable to adsorption since it provided higher cell retention. However, in the case of 11α-hydroxylation, when the cell concentration inside the support was reduced by 50%, activity of the catalyst was maintained over three repeated runs, compared with only one run for a high cell concentration.

In spite of a rather short half-life, the system discussed is the first successful attempt to apply immobilized living cells for progesterone α-hydroxylation. Several approaches to improve catalyst performance can be suggested: these include using lipophilic prepolymer resins to increase substrate availability to the entrapped cells, immobilizing microorganisms other than *R. nigricans* ATCC 6227b with more stable 11α-hydroxylase enzymes, utilizing various solvents, and finding different methods for NADPH regeneration as well as enzyme stabilization and reactivation.

7.4 11β-hydroxylation

Incorporating a hydroxyl group into the 11β-position of a steroid molecule by a microbial catalyst was accomplished by Larsson and Mosbach. This reaction could be performed by immobilized cells or immobilized spores.[5–7]

7.4.1 11β-hydroxylation by Immobilized Cells[5,6]

The reaction rate of *Curvularia lunata* ATCC 12027 cells immobilized into acrylamide gel was compared under two different modes of operation: batchwise and continuous column. Product formation by attached cells in shake flasks (28°C) was 2.0–2.9μM/hour/gr wet gel. 100% conversion was obtained in 15 hours using an initial substrate concentration of 7.2μM. Raising the substrate level to 28.8μM increased the time necessary for 100% conversion to 28 hours.

A column reactor packed with granules of *Curvularia lunata* was applied for substrate (.3μM) conversion. Although the initial transformation capacity was 0.15μM/hr/gr gel, it was shown that the initial conversion rate could be incresed to 0.5μM/hr/gr gel by reincubating the granules in a nutrient medium. Reactivation was accomplished as a result of cell multiplication inside the gel.

As a whole, 11β-hydroxylation by immobilized cells was subject to two limitations: low rate of product formation, and inhibition by high substrate concentrations (above 0.21 mg/ml).[5,6]

7.4.2 *11β-hydroxylation by Immobilized Spores*[7]

The immobilized spore approach was investigated as an alternative to steroid hydroxylation by attached cells.

Spores of *Curvularia lunata* ATCC 12027 were entrapped into acrylamide or calcium alginate gels and used for transformation of cortexolone to cortisone. Immobilized spores were first allowed to germinate inside the support for 24–72 hours at 28°C by placing the catalyst into a nutrient medium. The rate of spore germination inside the matrix depended on the type of carrier as well as the aeration rate. Germination proceeded for 24 hours in the calcium alginate gel with aeration, whereas under the same conditions germination in acrylamide took longer (up to 72 hours). The substrate transformation rate was also influenced by the type of support. The catalyst formed by spore entrapment into calcium alginate gel exhibited much higher product formation (43%) than spores immobilized into acyrlamide (2%).

The properties of free and alginate-entrapped spores were compared: Temperature optima as well as the rates of hydroxylation were the same, but immobilized cells exhibited higher temperature stability and a broader pH optimum. To increase substrate solubility, the effects of various solvents were investigated. Methanol (2%), dimethylsulfoxide (1%), Tween 80 (1%), and ethanol (1%) all increased substrate solubility and product yield, but raising the methanol level from 2% to 10% lowered transformation capacity of the immobilized spores. Also, an ethanol concentration of only 2% decreased cortisone yield. Entrapped spores exhibited a high storage stability (four months at 4°C), whereas immobilized mycelium was inactivated after only a few days of storage at the same temperature.

Transformation of cortexolone to cortisone by immobilized spores was performed in repeated batch operations at two different substrate concentrations. Using a high substrate concentration (1mM), catalyst activity declined after reincubation in a nutrient medium, whereas at a low substrate level (0.1mM), the transformation capacity of the carrier increased. This result points to inhibition of 11β-hydroxylase by high substrate concentrations and demonstrates tht pseudocrystallo-fermentation would be unfeasible for the process. However, the greater enzymatic activity of a reactivated catalyst after application at a low cortexolone concentration strongly suggests the feasibility of a continuous flow process using immobilized spores.

7.5   Side-chain Cleavage at the C_{17}-position

Side-chain cleavage of sterol molecules at the C_{17}-position without degrading the steroid nucleus is of great importance to industry since it allows the use of inexpensive plant sterols (cholesterol, β-sitosterol, campesterol) as raw materials for drugs. Currently available commercial processes are based on diosgenin, a phytosterol extracted from plants grown in Central America; however, supply from this area is diminishing.[28–31]

The sequence of reactions involved in side-chain cleavage of sterols is shown in Figure 1. This multi-step enzymatic process can be carried out by immobilized cells under various conditions; in the presence of inhibitors preventing steroid ring degradation, and in the presence of modified substrates.

Catalyst performance under various modes of operation will be discussed in detail.

7.5.1   Side-chain Cleavage in the Presence of Inhibitors[19b]

Various compounds are known to inhibit steroid ring degradation, including 8-hydroxychinoline, $Co(NO_3)_2$, $NiSO_4$, Pb and Se ions, 2,2'-dipyridyl-1,10-phenanthroline, etc.

In the presence of 8-hydroxychinoline, free *Nocardia erythropolis* cells can

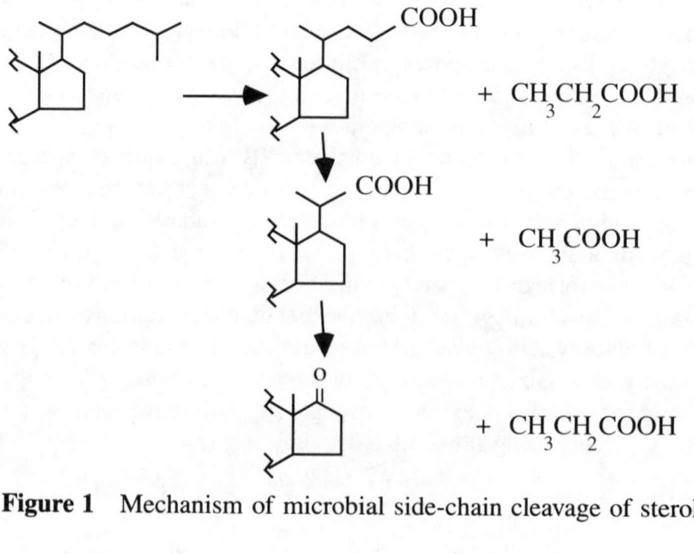

Figure 1   Mechanism of microbial side-chain cleavage of sterols.[9]

convert cholesterol to: I. cholestenone, II. a mixture of 20-carboxypregna-4-en-3-on and 20-carboxypregna-1,4-dien-3-on, III. a mixture of androst-4-en-3,17-dione and androst-1,4-dien-3,17-dione at a ratio of 10:70:20 (%). The transformation was carried out in 0.01M phosphate buffer (pH = 8.0) with an inhibitor concentration of 0.04 mg/ml. Also, substrate dissolved in dimethyl-formamide was added to the buffer at the concentration of 0.5 mg/ml.

Cells were immobilized into acrylamide, Fe^{+3}-hydroxigel, Al-alginate, Ca-alginate, and by adsorption to different types of celluloses, kieselgel, alumina oxide, etc.

In the presence of 8-hydroxychinoline, immobilized cells transformed cholesterol solely to cholestenon, with only one exception. The catalyst formed by microbial adsorption to cellulose powder converted cholesterol to a mixture of products I and II at the ratio of 30:50(%) (Table 9). When $Co(NO_3)_2$ or $NiSO_4$ were used instead of 8-hydroxychinoline, the transformation capacity of the matrix changed. Cells immobilized by adsorption to DEAE cellulose in the

Table 9   Side-chain Cleavage at the C_{17}-position by Immobilized Cells (Inhibitor: 8-hydroxychinoline)*

Supports and Methods of Immobilization	Products (%)		
	I	II	III
Free cells (control)	—	70	20
Entrapment:			
Acrylamide—15%	50	—	—
7.5%	60	—	—
5.0%	60	—	—
CNBr-activated Sepharose 4B	30	—	—
Fe^{+3}-hydroxygel	100	—	—
Fe^{+3}-hydroxygel/gelatin	100	—	—
Al-alginate gel	100	—	—
Ca-alginate gel	100	—	—
Adsorption:			
Cellulose powder	30	50	—
Microcrystalline cellulose (Chemapol)	70–90	—	—
Acetylcellulose	10	—	—
DEAE cellulose	35	—	—
Amberlite	60	—	—
Kiselgel (Chemapol)	20	—	—
Al_2O_3	60	—	—
Glasswool	—	—	—

*Microbe: *Nocardia erythropolis* IMET 7185; Substrate: Cholesterol; Products: I. Cholestenone; II. Mixture of 20-carboxypregna-4-en-3-on and 20-carboxypregna-1,4-dien-3-on; III. Mixture of androst-4-en-3,17-dione and androst-1,4-diene-3,17-dione presence of the inhibitor $Co(NO_3)_2$ formed mainly product II (70%), while exposure to $NiSO_4$ allowed synthesis of only product I (10%) (Table 10); the rest remained substrate. Using $Co(NO_3)_2$, cells entrapped into the Co/Al-Poly-acrylat/alginate support transformed cholesterol to a mixture of products I and II.

From the above results, it is clear that product formation can be regulated depending on the type of matrix and the inhibitor. This suggests that inhibitors either directly affect cell enzymes responsible for ring degradation, or they first react with the support matrix by forming a complex, freeing ions which block enzymes responsible for ring degradation.

Whatever the actual mechanism is, side-chain cleavage of cholesterol by attached cells can proceed in the presence of inhibitors, but special attention should be paid to carefully matching inhibitor and carrier properties.

7.5.2 Side-chain Cleavage of Structurally Modified Substrates

Selective side-chain cleavage of structurally modified substrates can also be performed by immobilized microbes. *Mycobacterium phlei* IMET SG 1026 and *Mycobacterium smegmatis* IMET H 124 cells were immobilized by entrapment into acrylamide, Al-alginate, or Co/Al-alginate gels and by adsorption to

Table 10   Side-chain Cleavage at the C_{17}-position in the Presence of $Co(No_3)_2$ or $NiSO_4$

Supports	Inhibitors	Products (%)		
		I	II	III
Polyacrylamide—7.5%	$Co(NO_3)_2$	70	+	—
Polyacrylamide—7.5%	$NiSO_4$	10	—	—
Co/Al-alginate/acrylamide	$Co(NO_3)_2$	50	+	—
Ni/Al-alginate/acrylamide	$NiSO_4$	30	—	—
DEAE cellulose/acrylamide	$Co(NO_3)_2$	90	—	—
DEAE cellulose/acrylamide	$NiSO_4$	+	—	—
Co/Al-alginate/DEAE	$Co(NO_3)_2$	+	50	10
Ni/Al-alginate/DEAE	$NiSO_4$	100	—	—
DEAE cellulose	$Co(NO_3)_2$	—	70	—
DEAE cellulose	$NiSO_4$	10	—	—
Co/Al polyacrylat	$Co(NO_3)_2$	10	10	—
Co/Al polyacrylat/alginate	$Co(NO_3)_2$	15	15	—
Co/Al polyacrylat/alginate/gelatine	$Co(NO_3)_2$	20	20	—

*Microbe: *Nocardia erythropolis* IMET 7185; Substrate: Cholesterol; Products: I. Cholestenone; II. Mixture of 20-carboxypregna-4-en-3-on and 20-carbodypregna-1,4-dien-3-on; III. Mixture of antrost-4-en-3,17-dione and androst-1,4-diene-3,17-dione cellulose. 3,3,-ethylenedioxy-5-cholesten or 4-cholesten-3-(O-carboxymethyl)-oxim (cholesterol derivatives) were used as substrates for conversion to the corresponding androstenedione derivatives. Transformation proceeded in 0.01M phosphate buffer (pH = 8.0), to which substrate dissolved in dimethylformamide was added. Conversion was performed batchwise at 37°C for 24 hours or as a repeated batch operation.

Final product formation depended on the microbial strain, nature of the support, method of immobilization, and the substrate utilized. Using 3,3-ethylenedioxy-5-cholesten, the catalyst formed by *M. smegmatis* adsorbed to DEAE-cellulose was more active (100%) compared to *M. phlei* microbes immobilized to the same support (45%), (Table 11). However, matrix productivity changed depending on the substrate: *M. phlei* cells adsorbed to DEAE-cellulose transformed 100% of 4-cholesten-3-(O-carboxymethyl)-oxim to product.

The most important parameter in system evaluation is its operational stability (Graphs A, C). Half-life of the catalyst was found to depend on the substrate and/or the type of support. The half-life of the *M. phlei* microbe-carrier complex used to ferment 3,3,-ethylenedioxy-5-cholesten was only two days and did not depend on the type of matrix. The same catalyst performed differently when 4-cholesten-3-(O-carboxymethyl)-oxim was used as substrate. Immobilization to acrylamide, DEAE-cellulose, and Al-alginate gel stabilized activity for a

Table 11   Side-chain Cleavage at the C$_{17}$-position on Modified Substrates

Microbe	Substrate	Method of Immobilization	Product (%)
Mycobacterium smegmatis IMET H-124	3,3-ethylene-dioxy-5-cholesten	DEAE, adsorption	100,* 10[†]
		Entrapment	
		Acrylamide—7.5%	traces
		Al-alginate	100,* 30[†]
		Co-Al alginate	30,* traces[†]
Mycobacterium phlei IMET SG 1026	3,3-ethylene-dioxy-5-cholesten	DEAE, adsorption	45*
		Entrapment	
		Acrylamide—7.5%	5
		Al-alginate	55*
	4-cholesten-3-(O-carboxymethyl)-oxim	DEAE, adsorption	100*
		Entrapment	
		Acrylamide—7.5%	75*
		Al-alginate	85*

Substrate: Cholesterol derivative; Product: Androstenedione derivative
*Batch cycle
[†]Repeated batch cycle

period of eleven days. During this time, conversion by cells adsorbed to DEAE-cellulose remained at 100% and 50% for Al-alginate entrapped cells. Microbes entrapped into acrylamide gel were stable for 11 days of operation at a transformation capacity of 70% (Graph A).

Several important changes were noticed in the properties of cells after immobilization: Attached cells exhibited a lower transformation capacity, had a wider temperature optimum (47–58°C for immobilized vs 47°C for free cells, Graph B) and were more stable in continuous operation. The half-life of free cells was shorter than 6 days, whereas immobilized cells were stable for more than 40 days (Graph C). The microbial catalyst could be reactivated by reincubation in a nutrient medium.

The results indicate that transformation of sterol derivatives to corresponding

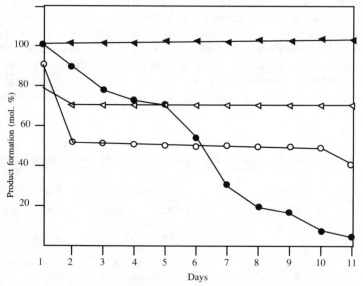

Transformation of cholesterol derivatives to androstenedione derivatives.[19c] Batchwise operation.

    ● —free cells;

    ◀ —DEAE cellulose adsorbed cells;

    ◁ —acrylamide gel entrapped cells;

    ○ —Al-alginate gell entraped cells.

Graph A   Side-chain cleavage at C_{17}-position: operational and temperature stability.

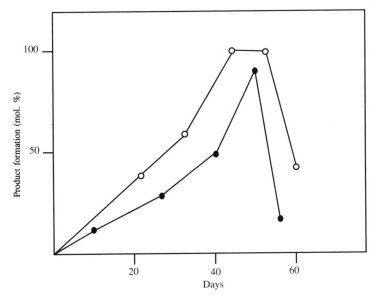

Changes in cell properties after immobilization: temperature stability.[20]
Microbe: *Mycobacterium phlei*;
Substrate: cholesterol derivative;
Product: androstenedione derivative;
      ●—free cells;
      ○—acrylamide entrapped cells.

Graph B

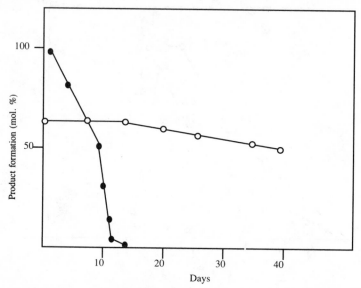

Transformation of cholesterol derivatives to androstenedione derivatives (continuous operation).[20]

● —free cells;

○ —acrylamide entrapped cells.

Graph C

androstenedione derivatives can be performed more cost effectively by immobilized cells than by free microbes. The excellent stability of cells adsorbed to DEAE-cellulose gives rise to the possibility of a continuous flow process using cholesterol derivatives.

7.6　16α-hydroxylation[35]

Incorporating a hydroxyl group into the 16α-position of a steroid molecule is also of major industrial importance. This reaction was recently accomplished with immobilized *Streptomyces roseochromogenes* cells, which converted dehydroepiandrosterone (DHEA) to 16α-hydro DHEA.

S. roseochromogenes cells were entrapped into two types of photocrosslinkable hydrophilic prepolymers (14 G and 46 G), differing in molecular weight (700 vs 2100). Transformation was performed with shaking (230 RPM) in Tris-HCl buffer (pH = 8.0, 27°C) at a substrate concentration of 2.1 × 10^{-4}M.

The hydroxylation rate by free cells gradually increased for the first 24 hours, giving a product yield of 80mM. The transformation capacity then leveled off for an additional 20 hours and ceased. Cells entrapped into both prepolymers also performed 16α-hydroxylation. However, by contrast with free microbes, the conversion rate gradually rose for 48–60 hours. The product yield was higher for 14 G than for 46 G entrapped cells (140mM vs 100mM).

The transformation capacities of catalysts greatly depended on the hydrophobicity of supports. Hydrophilic prepolymers (like 14 G and 46 G) generally exhibited 56–60% of the enzymatic activity of free cells, whereas hydrophobic resins showed no activity. When the co-factor $NADH_2$ was added to both systems, the reaction was carried out only by the hydrophilic carrier, indicating that the lipophilic matrix had no affinity for the co-factor.

No difference in the pH profiles of free and immobilized cells was observed; in both systems, 16α-hydroxylation peaked at a pH = 8.0. The enzymatic stability of attached microbes was much greater than for free cells, whose activity diminished after 24 hours. Catalyst half-life was 240 hours, which may be due to the stabilizing effect of the matrix.

One advantage of immobilized cells over immobilized enzymes is that the former can regenerate coenzymes. 16α-hydroxylation of DHEA specifically requires $NADH_2$. Adding this coenzyme to the microbial catalyst increased 16α-hydroxylation to 140mM, but attached cells could perform the reaction without $NADH_2$ at a productivity of 80–100mM.

The high yield and long half-life make the immobilized microbial system very attractive for a continuous flow 16α-hydroxylation process.

7.7 Summary

Applying microbial catalysts for steroid modifications represents the most recent alternative to the already existing technologies: chemical synthesis, free enzymes, immobilized enzymes, and free cells. The major advantage of immobilized microbes is that after entrapment into some support matrices, they remain alive, able to multiply and carry out single as well as multi-step enzymatic reactions. In addition, the catalysts can be reactivated.

Effectiveness of attached cells depends on properties of the strain, retention capacity of the support, the carrier contribution, method of immobilization, type of solvent, availability of oxygen, and the substrate-product influence on enzymatic activity of cells. Each component of the system contributes to catalyst performance and when properly matched, product yield and operational stability of the process may be higher than for free cells. Immobilized cells exhibited higher pH and temperature optima than free microbes and had longer operational stabilities.

Microbial catalysts were used to carry out the following steroid transformations: $11\alpha,11\beta,16-\alpha$-hydroxylation, Δ'dehydrogenation, and side-chain cleavage of sterols at the C_{17}-position. 11β-hydroxylation could be performed on a continuous flow basis by immobilized spores at low substrate concentrations. By contrast, steroid Δ'-dehydrogenation was more efficiently done by attached cells in repeated batch operations at high substrate levels. It was also shown that newly developed prepolymer resins with different hydrophobicities could be advantageous for immobilization. Side-chain cleavage of cholesterol was performed by attached cells on modified substrates and in the presence of inhibitors, via batchwise or continuous fermentation. In all cases, catalyst incubation in a nutrient medium reactivated the immobilized cells.

Depending on the specific transformation and oxygen requirements, one may prefer batch operation at high substrate levels to a continuous flow process using organic solvents. The application of newly developed prepolymer resins looks promising since it allows tailoring matrices for specific purposes and also increases substrate availability to entrapped cells. Most recent results show that due to cell-support interactions, toxic carriers (such as acrylamide) may inhibit unwanted enzymes (9α-hydroxylase) and can activate the enzymes responsible for side-chain cleavage of sterols.

Surprisingly, biochemically blocked mutants were not used for catalyst formation. However, applying them to specifically tailored supports would result in development of stable catalyst systems and may lead to industrial processes for steroid modifications using living immobilized microbial cells.

Literature

1. Klyne, W., *The Chemistry of the Sterols*, John Wiley & Sons, Inc. (1965)
2. Charney, W., Herzog, *Microbial Transformation of Sterols*, Academic Press, New York (1967)
3. Abbott, B., *Advances in Applied Microbiology*, **20**, 203 (1976)
4. Penasse, L., Peyre, M., *Sterols*, **12**, 525 (1968)
5. Mosbach, K., Larsson, P. O., *Biotech. Bioeng.*, **12**, 19 (1970)
6. Larsson, P. O., Mosbach, K., *Methods in Enzymology*, Mosbach, K. (Ed.), Vol. 44. 183 (1976)
7. Ohlson, S. A. et al., *Eur. J. Appl. Microb. Biotech.*, **10**, 1 (1980)
8. Sebek, O. K., Perlman, D., *Microbial Technology* Peppler, H. and Perlman, D. (Eds.), Academic Press, Vol. 1 (1979)
9. Schoemer, U., Martin, C. K. A., *Biotech. Bioeng.*, **22**, Supp. I, II (1980)
10. McGregor, et al., *Biotech. Bioeng.*, **14**, 831 (1972)
11. Larsson, P. O., et al., *Nature*, **263**, 796 (1976)
12. Larsson, P. O., et al., US Patent 4,246,346 (1981)
13. Ohlson, S. A., et al., *Biotech. Bioeng.*, **20**, 1267 (1978)
14. Venkatasubramanian, K., et al., *Enzyme Engineering*, Rey, E. K., Weetall, H. H. (Eds.), Vol. 3, 29 (1979)

15. Constantinidis, A., *Biotech. Bioeng.*, **22**, 119 (1980)
16. Koshcheenko, *et al.*, *Applied Biochem. Microb.*, **10**, 3, 376 (1974)
17. Koshcheenko, *et al.*, *Eur. J. Appl. Microb. Biotech.*, **12**, 3, 161 (1981)
18. Yang, H. S., Studebaker, J. F., *Biotech. Bioeng.*, **20**, 17 (1978)
19. Atrat, P. *et al.*, *Zeitschrift fur Allgemeine Microb.*, **20**, 2, 79 (1980a); *Z. fur Allgem. Microb.*, **20**, 3, 159 (1980b); *Z. fur Allgem. Microb.* **20**, 4, 239 (1980c)
20. Atrat, P. *et al.*, *Eur. J. Appl. Microb. Biotech.*, **12**, 157 (1981)
21. Omata, *et al.*, *Eur. J. Appl. Microb. Biotech.*, **6**, 207 (1979)
22. Fukui, *et al.*, *Eur. J. Appl. Microb. Biotech.* **10**, 289 (1980)
23. Sonomoto, K., *et al.*, *Agricult. Biolog. Chemistry*, **44**, 5, 1119 (1980)
24. Ohlson, S. A., *et al.*, *Eur. J. Appl. Microb. Biotech.* **7**, 203 (1979)
25. Arinbasarova, A., *et al.*, *Applied Biochem. Microb.*, **16**, 5, 854 (1980)
26. Koshcheenko, *et al.*, *Applied Biochem. Microb.*, **12**, 2, 206 (1976)
27. Kondo, E., Masuo, E., *J. Applied Microb.*, **7**, 2, 113, 1961
28. Marsheck, W. J., *Progress in Industrial Microb.*, **10**, 49 (1971)
29. Marsheck, W. J., *J. Applied Microb.*, **23**, 1, 72 (1972)
30. Jin, J., Marsheck, W. J., US Patent 3,994,933, (1976)
31. Jin, J., Marsheck, W. J., US Patent 4,032,408, (1977)
32. Lee, B. K., *et al.*, *Biotech. Bioeng.*, **13**, 503, (1971)
33. Kolot, F. B., *Proc. Biochem.*, **16**, 6, 30 (1981a); *Proc. Biochem.*, **16**, 5, 2 (1981b); *Proc. Biochem.*, **15**, 7, 2 (1980c); *Developments in Industrial Microbiology*, **21**, 29 (1980d)
34. Maddox, *et al.*, *Biotech. Bioeng.*, **23**, 345 (1981)
35. Chun, Y. Y. *et al.*, *J. Gen. Appl. Microb.*, **27**, 505 (1981)
36. Okuno, H., *J. Appl. Biochem.*, **7**, 3, 192 (1985)
37. Janolino, G., *J. Appl. Biochem.*, **7**, 1, 33 (1985)
38. Kumakura, M., *Biosci. Rep.*, **4**, 3, 181 (1984)
39. Tysiachnaia, I., *Prikl. Biokhim. Mikrobiol.*, **20**, 1, 79 (1984)
40. Verani, J., *Biotech. Bioeng.*, **25**, 5, 1359 (1983)

8

Microbial Systems for Amino Acids Production

The eighth chapter introduces industrial and laboratory scale applications of immobilized cells for amino acid synthesis. For each process, the method of attachment, half-life, temperature stability, and type of reactor utilized are discussed. The effects of particle size, as well as cell and substrate concentrations on the reaction rates are presented. Throughout the chapter, the advantages of immobilized cells over free cells and immobilized enzymes are emphasized. Although it was confirmed that multistep enzymatic reactions can be carried out using microbial catalysts, the specific rates of product formation during the transformations were usually rather low. By contrast, one step enzymatic processes could be performed with high yields.

8.0   L-aspartic Acid Production

Aspartic acid is used in medicines, as a food additive, and as the starting material for synthesis of the sweetener aspartame. Conversion of fumaric acid to aspartic acid is a one step enzymatic reaction:

$$HOOCCH = CHCOOH \xrightarrow[\substack{\text{aspartase} \\ \text{(E.C.4.3.1.1.)}}]{+ NH_4^+} HOOCCH_2 - \underset{\underset{NH_2}{|}}{CHCOOH}$$

fumaric acid   aspartase   L - aspartic acid
                     (E.C.4.3.1.1.)

The aspartase enzyme catalyzes addition of an ammonium ion to the double bond of fumaric acid, forming L-aspartic acid.

The reaction can be performed as a batch process using free cells (free enzymes) or as a continuous process using immobilized cells. The free enzyme system has three disadvantages:

1. Loss of enzyme activity as a result of enzyme extraction from cells (aspartase is an intracellular enzyme).
2. Instability of the free enzyme.
3. Periodicity of the process.

The immobilized enzyme system, however, showed low activity (29% with acrylamide gel and 46% with carrageenan gel used as support matrices) and poor operational stability (20 days at 37°C). The immobilized enzyme system was thus deemed unsatisfactory for industrial application, and the immobilized cell approach was tested. The result: A very effective immobilized cell system, industrialized since 1973 by the Tanabe Seiyaku Co. in Japan. This system is described below.

E. coli ATCC 11303 cells were grown under aerobic conditions for 24 hours at 37°C on a medium containing corn steep liquor (4%), ammonium fumarate (3%), and salts. Immobilization was then carried out by entrapment into acrylamide gel. Gel particles, preincubated 48 hours at 37°C with 1M ammonium fumarate (pH = 8.5) containing 1mM Mg^{+2}, were packed into a 2.2×9.0 cm column. Initially, a solution of 1M ammonium fumarate (pH = 8.5, 1mM Mg^{+2}) was continuously pumped through the column at the flow rate of SV (space velocity) = 1.5 hr^{-1}. The following factors affected L-aspartic acid production:

1) *Temperature*. The optimal temperature for 100% product formation was

Table 1

Disadvantages of free and immobilized enzyme systems
1. Enzyme isolation and purification results in loss of enzymatic activity
2. An expensive electron acceptor is needed throughout transformation, and must often be renewed
3. The immobilized enzyme system does not retain sufficient activity

Disadvantages of the free cell system
1. Low operational stability
2. Cells cannot be reactivated
3. Substrate/product inhibition in the batchwise process
4. Periodicity of the transformations
5. Low heat stability

45–50°C. Increasing the temperature to 55°C reduced aspartic acid production to 75–80%.

2) *pH*. The maximum relative activity of the immobilized cell catalyst occurred at a pH of 8.0.

3) *Metal ions*. Mg^{+2}, Mn^{+2}, Ca^{+2}, ions at the concentration of 1mM were absolutely essential for stable column operation. In the presence of metal ions, 100% substrate formation was observed after 16 days of operation. Without metal ions, the half-life of the system dropped to only 14 days, while product formation decreased simultaneously (ceteris paribus). Authors contributed the stimulatory effect of metal ions to their ability to protect the cells against heat liberated during polymerization. Another explanation could be that metals activate the aspartase enzyme itself.

4) *Medium flow rate*. One hundred percent product formation was observed only at flow rates $SV < 0.7 \ hr^{-1}$.

5) *Operational stability of the immobilized cell column*. The operational stability of the immobilized cell column depended on the medium flow rate. At $SV = 0.5 \ hr^{-1}$, 100% product formation was observed during 40 days of operation. Increasing the flow rate to $SV = 0.8–1.2 \ hr^{-1}$ reduced product formation to 70–80%.

Allowing immobilized cells to stand in a substrate solution for 24–48 hours at 37°C increased activity by a factor of 10. This increase was observed in the presence of chloramphenicol, an inhibitor of protein synthesis. The observed activation was not attributed to the protein synthesis, but rather to autholysis of *E. coli* cells entrapped into the acrylamide gel matrix, which was later confirmed by electron microscopy observation. It thus appeared that lysis occurred upon cell entrapment into acrylamide, but the aspartase enzyme could not leak from the matrix, whereas the substrate and product could easily pass through it.

A solution of 1M ammonium fumarate (pH = 8.5, 1mM Mg^{+2}) was passed through an immobilized cell column (100 × 10 cm) at the flow rate of 6 liters/hour and at 37°C. L-aspartic acid was recovered from the effluent, giving a product yield of 95%. The half-life of the system was 4 months at 37°C. Compared with the soluble enzyme process, a 60% reduction in cost was achieved. Recently, performance of the immobilized cell system was significantly increased by substituting the K-carrageenan matrix for the acrylamide support. Aspartase activity, hardening treatment, and half-life of the K-carrageenan system are presented in Table 2.

Enzymatic activity of the K-carrageenan microbial catalyst was significantly higher than that of acrylamide entrapped cells (56,340 vs 18,850), but the stability of the preparation was relatively low (70 days vs 120 days). To increase the half-life of the system, hardening treatments with various concentrations of hexamethylenediamide and glutaraldehyde were carried out. Although treat-

Table 2   Hardening Treatment for Stabilizing Aspartase Enzyme Activity of *E. coli* Immobilized into K-carrageenan Gel

Hardening Treatment			Aspartase Activity after Activation (unit/gr. cells)	Half-life at 37°C (days)
Reagent and Final conc.	Temperature (°C)	Time of Treatment		
None			56,340	70
Persimmon tannin	37°	2 hr.	54,500	86
GA:* (mM)	4°	15 min.		
2.5			39,690	113
5.0			37,460	240
10.0			28,040	252
HMDA*-GA:	4°	30 min.		
(mM) (mM)				
85.0 1.7			50,210	75
85.0 17.0			49,900	108
85.0 85.0			49,400	680
8.5 17.0			36,120	443
Polyacrylamide (control)			18,850	120

*GA: Glutaraldehyde; HMDA: Hexamethylenediamide

ment with hardening agents lowered enzymatic activity of the immobilized cell preparation, it significantly increased the operational stability of the system. Using 85mM HMDA and 85mM GA, the half-life of the system was raised to 680 days.

Productivities of the two *E. coli* systems were calculated from the equation:

$$\int_{0}^{t} E_o \exp\,(-k_d \cdot t)\; dt$$

E_o = initial activity; k_d = decay constant; t = operational period. The relative productivity of the immobilized cell preparation formed by cell entrapment into K-carrageenan gel was taken as 100. The productivity of the immobilized cell preparation formed by cell entrapment into K-carrageenan gel and hardened with HMDA and GA was determined to be 15 times higher, while the operational stability was 5.5 times greater.

As can be concluded, the application of the K-carrageenan gel is extremely successful for industrial production of L-aspartic acid.

Literature

Zueva, N., *Prikl. Biokhim.*, **21**, 3, 333 (1985)
Umemura, I. *et al.*, *Appl. Micro. & Biotech.*, **20**, 5, 291 (1984)
Fusee, M. *et al.*, *Appl. Env. Micro.*, **42**, 672 (1981)
Takamutsu, S. *et al.*, *J. Ferm. Tech.*, **58**, 2, 129 (1980)
Chibata, I., *ACS Symposium*, Series 106, 13 (1979)
Sato, T. *et al.*, *Biochem. Biophs.*, **570**, 179 (1978)

8.1 L-alanine Production

Alanine is an essential amino acid which plays an important role in protein, nucleic acid and lipid metabolism. It can be converted to pyruvate, a key intermediate for synthesis of the above mentioned compounds, via NAD-dependent alanine dehydrogenase. By the action of another enzyme, transaminase, L-alanine can also be transformed to L-glutamate, L-valine, L-leucine, and L-phenylalanine (Fig. 1).

Several methods to synthesize L-alanine are described in the available literature. Production by multi-step enzymatic reactions of various microorganisms (*Bacillus subtilis, B. brevis, Escherichia coli, Clostridium kluyver, Mycobacterium album, Corynebacterium, Streptomyces species,* etc.) is presented, and strain isolation from various sources, selection of highly active mutants, the possible mechanisms of alanine synthesis from different carbon/nitrogen sources, as well as the enzymes involved are described. Another method presented is formation of L-alanine from L-aspartic acid via a batchwise one step enzymatic reaction using L-aspartate-β-decarboxylase enzyme (E.C. 4.1.1.12) and requiring the cofactor pyrodoxal phosphate (Fig. 2).

However, the two most interesting approaches to enzymatic conversion of

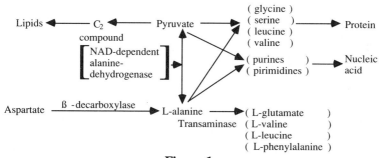

Figure 1

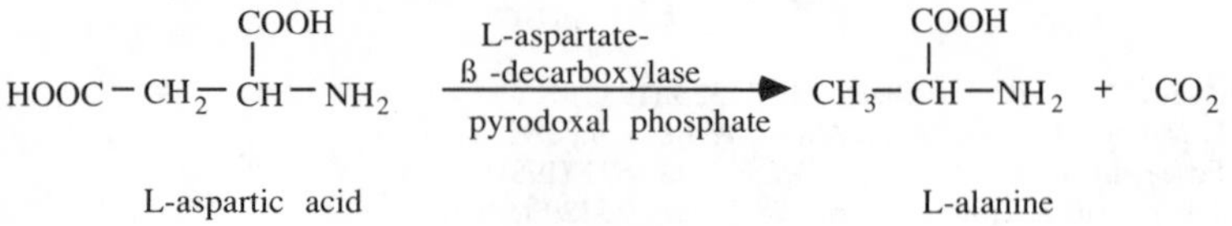

$$\underset{\text{L-aspartic acid}}{HOOC - CH_2 - \underset{\overset{|}{COOH}}{CH} - NH_2} \xrightarrow[\text{pyrodoxal phosphate}]{\substack{\text{L-aspartate-} \\ \text{β -decarboxylase}}} \underset{\text{L-alanine}}{CH_3 - \underset{\overset{|}{COOH}}{CH} - NH_2} + CO_2$$

Figure 2

L-aspartic acid to L-alanine are the free-cell batchwise process, and the continuous immobilized cell process; both are outlined below.

L-aspartate-β-decarboxylase was found in a number of microorganisms, such as *Pseudomycobacterium, Clostridium perfringens, Desulfovibrio desulfuricans, Nocardia globerula, Pseudomonas reptilivora, Acetobacter species, Achromobacter, Alcaligenes falcalis, Xanthomonas oryzae*, and *Pseudomonas docunhae*. The *P. docunhae* IAM 1152 strain was the most active organism among the numerous microbes tested.

8.1.1 Transformation by Free P. docunhae Cells in a Batchwise Process

P. docunhae cells cannot utilize sugars for growth, so a monobasic organic acid such as fumarate was the most effective carbon source. Various other monobasic, dibasic, tribasic, and amino acids were also tried, but it appeared that sodium glutamate at a concentration of 2% gave the highest total and specific enzyme activity (9.2 units/ml and 2.2 units/ml, respectively). L-glutamate stimulated β-decarboxylase formation in all microorganisms tested, but the highest activity was again recorded by the *P. docunhae* strain IAM 1152 (Table 3).

100 ml of *P. docunhae* cells were added to a medium (37°C) containing L-glutamate (2%), L-aspartic acid (100gr), and 7 ml of 28% NH_4OH. The amount of L-alanine increased linearly with the consumption of L-aspartic acid. Initially, the substrate concentration exceeded its solubility (pseudocrystallo fermentation), but undissolved crystals solubilized into the solution during the transformation. When L-alanine accumulated in excess of its solubility, crystals percipitated from solution, yielding 66 gr/100ml, or 93%. Neither formation of other acids nor decomposition of L-alanine occurred. During the reaction, the pH was maintained constant at 5.3, optimal for enzyme activity, but it rapidly rose to 9.2 at the conclusion of the reaction. Charcoal was then added to the alanine in the reaction mixture, the pH was adjusted to 4.0 and the mixture was boiled and filtered at T > 60°C. The solution formed was mixed with an equal amount of methanol and cooled to +4°C. Crystalline L-alanine was collected and recrystallized.

Table 3 Effect of L-glutamate on L-aspartate-β-decarboxylase Formation by Various Microorganisms

Microorganism	Carbon Source	Growth ($A_{600} \times 20$)	Enzyme Activity	
			Total (u/ml broth)	Specific (u/mg protein)
P. docunhae	Fumarate	0.305	4.02	1.0
IAM 1152	L-glutamate	0.340	7.25	1.95
A. liquidum	Fumarate	0.106	0.102	0.085
IAM 1446	L-glutamate	0.127	0.185	0.152
A. pestifer	Fumarate	0.347	0.255	0.082
IAM 1667	L-glutamate	0.365	0.520	0.172

8.1.2 Continuous Transformation in a Column Reactor by Immobilized Cells

Continuous production of L-alanine from L-aspartic acid was demonstrated using *P. docunhae* IAM 1152 cells immobilized by entrapment into the K-carrageenan gel.

The gel was first dissolved in a 0.9% NaCl solution (45°C), and added to the cells. The mixture then cooled to 5°C, forming a gel and thus entrapping the microbes. In order to strengthen the gel, it was soaked in a 2% KCl solution at 5°C for 16 hours, after which the support was granulated into 3mm diameter particles and further hardened by glutaraldehyde treatment (0.2% for 10 min. at 10°C).

The potentially toxic effect of glutaraldehyde on cells was investigated. Relative activity of free cells was compared to untreated and glutaraldehyde treated matrices: The relative activity of a microbial catalyst formed with treatment was four times higher compared to free cells. This may be because the cell membrane presents a barrier to substrate-product diffusion, while glutaraldehyde treatment enhances membrane permeability. With immobilization per se, 89% of the substrate was converted to L-alanine; glutaraldehyde treatment reduced the relative activity to 78%. Substrate conversion by carrageenan entrapped cells after the first cell reactivation was only 26%, but increased following the second reactivation (44hr) to 59% and remained stable at that level until 68hr. By contrast, after the first reactivation, the treated matrix was two times more active (52%) than the untreated matrix. At 40 hr, it was only slightly less active (by 6–7%) compared with the non-treated matrix.

These results show that chemical treatment of the cell catalyst provides a support with high mechanical strength, good operational stability, yet gentle enough to be used with living cells.

The rather low matrix activity after reactivation (26–59%) may suggest the existence of 1. a diffusion barrier for substrate/product, due to the support matrix, cell membrane, or both 2. cell reactivation was not completed due to absence of carbon and nitrogen sources in the reactivating medium, causing reactivation of only the β-decarboxylase enzyme but not the entire cell.

The L-aspartate β-decarboxylase enzyme requires the cofactor pyrodoxal phosphate (PLP), addition of which to the medium at a concentration of 0.1–1 mM resulted in excellent enzymatic activity of the system (96–147%) during repeated operation.

The optimal pH range for treated and untreated matrices was from 5.5–6.5 units, whereas the free cell system exhibited maximum activity only at a pH of 6.2 units. Optimal temperature for enzymatic activity of free cells was found to be 53°C; immobilized cells preferred 57°C.

A glutaraldehyde treated matrix (12.6 gr) was packed into a column (1.6 × 12 cm) and 2M ammonium L-aspartate (pH = 6.2) containing 0.1 mM PLP was passed through the column at 57°C. With a retention time of 8 hours, almost 100% conversion occurred. An untreated matrix exhibited maximum relative activity (90%) between 10 and 20 days of operation, but conversion drastically diminished, reaching only 20% after 40 days. By contrast, the treated matrix was only slightly less active during the first 10–20 days of operation (85% relative activity), but was much more stable over the long run (half-life = 46 days).

Literature

Mosbach, K., *Ann. NY Acad. Sci.*, **434**, 239 (1984)
Sonomoto, K. et al., *Agr. Biol. Chem.*, **44**, 5, 1119 (1980)
Yamamota, K. et al., *Biotech. Bioeng.*, **22**, 2045 (1980)
Shibatani, T. et al., *Appl. Envr. Micro.*, **38**, 3, 359 (1979)

8.2   L-isoleucine Production

L-isoleucine is presently produced from glucose and D-threonine by batch fermentation using *Serratia marcescens* AT 130-1 cells. The process is a multi-step enzymatic reaction catalyzed by different enzymes and requires ATP and co-enzyme regeneration. Besides *S. Marcescens* AT 130-1 cells, other microorganisms such as *Brevibacterium flavum* and *M. parafinolyticus* can synthesize this amino acid from acetic acid, β-paraffins, etc.

As an alternative to the traditional fermentation process, application of immobilized living microbial cells was recently investigated.

Serratia marcescens AT 130-1 cells were immobilized by entrapment into

K-carrageenan with an initial concentration of 10^7 cells/ml. Gel beads (4 mm diameter) were granulated and incubated in a complete medium (glucose- 2%, dextrin- 4%, urea- 0.5%, D-threonine-1.5%, and salts) for 32 hours at 30°C. During this time, the cell concentration inside the gel increased from 10^7 to 10^{10} cells/ml. Interestingly, the concentration of living cells inside the gel and the amount of L-isoleucine produced (0.7–0.8 mg/ml/hr) were equal for free and immobilized microbes. However, immobilized microbes showed a ten fold lower oxygen demand compared with free cells (10^{-4} moles/ml/hr for free cells vs 10^{-5} moles/ml/hr for immobilized cells).

Although the need for oxygen fell due to entrapment, attached cells still required some for cell maintenance inside the matrix, thus dictating the reactor choice. In fact, comparing the column and fluidized bed reactors showed that only the later performed satisfactorily. The concentration of living cells in the column reactor decreased from 10^{10} cells/ml to 10^7 cells/ml, oxygen demand decreased from 10^{-5} moles/ml/hr to 10^{-7} moles/ml/hr, and L-isoleucine production diminished to zero during the first 48 hours of operation. This result indicates that oxygen is essential for successful continuous operation using immobilized living *S. marcescens* cells.

Effect for D-threonine on L-isoleucine Production

D-threonine is the L-isoleucine precursor, thus its quantity in the medium is vital. For a free cell system, a linear relationship between the D-threonine concentration in the range of 1–2% and the rate of L-isoleucine production was observed. In the immobilized cell system, this relation was non-linear; maximum L-isoleucine production (3 mg/ml) occurred at a 1.5% D-threonine concentration.

Effect of Sugar Sources on L-isoleucine Production

Raising the glucose concentration in the medium from 1% to 4% increased product synthesis. But the combination of rapidly utilized glucose (2%) and slowly utilized dextrin (4%) resulted in the highest yield (3.0 mg/ml/hr). At these sugar concentrations, L-isoleucine was produced at a constant rate for a 10 hour period.

Other important factors in L-isoleucine synthesis by immobilized cells were pH (maximum production was observed at a pH of 7.5) and a 1:2 volume ratio of immobilized gel particles:total reactor volume (50 ml of gel:100 ml fluidized bed volume).

Continuous L-isoleucine fermentation was carried out with two fluidized bed reactors placed in a single stream. The concentrations of living cells inside the matrices of both reactors were equal (5×10^{10} cells/ml) during 15 days of operation, and in both vessels glucose was fully consumed. The rate of total product formation was constant for more than 15 days; closer inspection revealed that product yield in the first reactor was 2.4 mg/ml, 1.5 times lower than in the second bed (3.6 mg/ml). Since cells in the second bed were relatively deficient in sugar, a mixture of glucose and dextrin (24% glucose, 48% dextrin) was constantly passed through it. This increased the amount of isoleucine produced in the second reactor to 4.5 mg/ml, and the system operated continuously at that level for more than 30 days.

It was observed that cells were continuously released from the matrix at a constant rate, but because *S. marcescens* cells multiplied inside the support, the concentration of living cells inside the gel remained the same.

The author concluded that the disadvantage of this system was the oxygen limiting conditions under which the product was formed. If oxygen transport into the gel could be increased, L-isoleucine yield would also rise and the operation as a whole could compete with presently existing batch fermentation.

Literature

LeBlond, D., Wood, W., *Arch. Biochem. & Biophys.*, **239**, 2, 420 (1985)
Grote, W. *et al.*, *Biotech. Letters*, **2**, 11, 481 (1980)
Wada, *et al.*, *Biotech. Bioeng.*, **22**, 6, 1175 (1980)

8.3   L-tryptophan Production

L-tryptophan is an essential amino acid generally lacking in cereal proteins. Tryptophan deficiency is observed in patients with diabetes, tuberculosis, and cancer. Tryptophan inadequacy in food products results in various forms of vitamin deficiencies.

Tryptophan can be obtained by protein hydrolysis, chemical synthesis, or microbial biosynthesis. During protein hydrolysis, a mixture of various amino acids is formed and difficulties are encountered in their separation. Chemical synthesis using acrylic acid as the starting material requires five subsequent steps, and DL-tryptophan is also formed. Separating L-tryptophan from the DL mixture presents difficulties, although a method is known resolving L-amino acids from acetyl-DL-amino acids by employing the immobilized aminoacylase enzyme system. Microbial conversion can be performed by a free cell batchwise operation with specifically selected microbes such as *Hansenula, Candida, Torula, Micrococcus luteus* ATCC 398, *M. lysodeikticus* B-61-2, *Bacillus megaterium, Serratia, E. coli,* etc. However, precursors like indole or antramilic acid are required for high product yield. Development of a continuous process using immobilized living microbial cells thus represents a novel approach.

8.3.1 Formation of L-tryptophan from Indole

Formation of tryptophan from indole is catalyzed by the enzyme tryptophan synthetase and by pyrodoxal phosphate.

indole + sodium pyruvate + ammonium bicarbonate ────────────▶ tryptophan

tryptophan synthetase
+ pyrodoxal phosphate

Continuous microbial fermentation was demonstrated using the *E. coli* K_{12} strain immobilized by entrapment into acrylamide gel. L-tryptophan synthetase activity in the cells was induced by adding 1 mg/ml of tryptophan to the fermentation medium seven hours prior to cell immobilization. Indole, sodium pyruvate, and ammonium bicarbonate were used as substrates. Product yield increased by 50% after addition of the cofactor pyrodoxal phosphate.

The reaction was performed in a continuously stirred reactor (CSTR) at 60 RPM; and in a plug flow reactor (both at 37°C, pH = 8.8–9.0). Apparently, the critical factor in obtaining an active microbial catalyst is acrylamide monomer content of the support. Increasing the amount of acrylamide from 500 mg to 1000 mg resulted in a three fold lower tryptophan yield (from 37 to 12μmol/hr/100 gr cells). This confirms acrylamide toxicity to the cells.

N-methylene bis acrylamide (BIS) concentrations in the range of 2–10% had no effect on system performance, but the stability and mechanical properties of the gel deteriorated considerably with BIS concentrations below 2.5%.

Microbial density inside the gel significantly influenced catalyst performance. Matrix activity increased linearly with increasing cell mass inside the gel; maximum product synthesis was observed with 400–450 mg bacteria per gram of gel.

Compared with the free cell system, enzymatic properties of the microbes did not change after attachment. The pH profiles for entrapped and free cell preparations were similar, with maximum activity occurring at pH = 9.0.

Maximum product conversion (100%) by the biocatalyst was observed between 35–40°C, but drastically diminished to less than 40% at 45°C.

The immobilized cell preparation was unstable in phosphate buffer (0.04M, pH = 8.0, 37°C). After 3 days, only 10% of the initial activity remained; however, when the medium fed to the entrapped microbes included all substrates and pyrodoxal phosphate, 70% of tryptophan synthetase activity remained after 5 days.

In both the free and immobilized cell systems, maximum product synthesis (27μmoles/100 mg cells) was observed at a 1M sodium pyruvate concentration.

The amount of indole in the medium was critically important for fermentation.

At the optimal cell count inside the gel and SV = 2.5 hr^{-1}, an increase in indole concentration from 7mM to 15mM resulted in 80% tryptophan production. Indole levels above 15mM caused a rapid deterioration in product formation (10%).

Continuous L-tryptophan synthesis was carried out in a plug flow reactor containing 3 gr of entrapped cells (indole concentration = 10 mM, pH = 9.1, 37°C). Because sodium pyruvate and ammonium bicarbonate substrates were present in much higher concentrations than indole (which was fully utilized), they were recycled by passing the mixture of tryptophan and unused substrates through a column filled with activated charcoal. Tryptophan was retained on the support, while sodium pyruvate and ammonium bicarbonate were mixed with indole and returned to the reactor.

The initial conversion rate was 80%, but product formation decreased with time. The estimated half life of the system was 10 days. L-tryptophan was recovered from the effluent with an approximately 50% yield.

Reasons for the declining productivity were not investigated. They may be a reduction in activity of entrapped cells or inhibition by products of a pyruvate alkaline reaction, which accumulate in the medium. Cell reactivation in the immobilized state was also not performed, although this could have increased product yield and system half-life. Despite minor deficiencies, the system as a whole is extremely interesting since it represents an attempt to continuously produce tryptophan via an immobilized cell catalyst.

Genetic manipulation techniques resulted in reconstruction of the *E. coli* strain and formation of multiple-mutants. Multiple-mutation strains were obtained with defined genotypes, capable of converting cheap carbon and nitrogen sources such as glucose and amino salts into tryptophan with high product yields (80mg/gr dry weight/hr).

Entrapment of these highly producing mutants into proper matrices may form extremely active catalysts which could carry out a continuous flow process for L-tryptophan fermentation.

Literature

Powell, L., *Biotechnol. Genet. Eng. Rev.*, **1**, 1 (1984)
Bang, W., Behrendt, U., *Biotech. Bioeng.*, **25**, 4, 1013 (1983)
Chua, J. *et al.*, *J. Ferm. Tech.*, **58**, 2, 123 (1980)
Tribe, D., Pittard, J., *Appl. Envr. Micro.*, **38**, 2, 181 (1979)
Decottignies, P. *et al.*, *Eur. J. Appl. Micro. & Biotech.*, **7**, 33 (1979)

8.4   L-citrulline Production

Conversion of L-arginine to L-citrulline is a one step enzymatic reaction carried out by the L-arginine deiminase enzyme (Fig. 3).

This transformation can be performed by free L-arginine deiminase producing microorganisms, which are cultivated in a nutrient medium containing L-arginine, or by the free L-arginine deiminase enzyme, which is extracted from the microorganisms and then reacted with the substrate. Both methods are disadvantageous for commercial production of L-citrulline because additional steps are required for removing the product from a medium contaminated with enzymes or microbial cells, and nutrients. In addition, batchwise operation does not allow reusing the free cells or enzymes.

Several microorganisms with L-arginine deiminase enzyme activity were immobilized by entrapment into acrylamide gel. They included *S. faecalis* ATCC 8043, *Pseudomonas putida* ATCC 4359, *Pseudomonas fluorescens* IFO 3081, *Pediococcus cerevisiae* P60 ATCC 8042, *Sarcina luted* IAM 1099, *Mycobacterium avium* IFO 3154, *Streptococcus grisems* IFO 3122, and *Streptococcus faecalis* ATCC 11420.

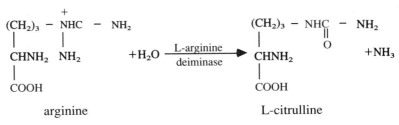

arginine   L-citrulline

Figure 3

The *Streptococcus faecalis* ATCC 8043 strain transformed arginine all the way to putrescine via the following multi-step enzymatic reaction:

$$\text{arginine} \xrightarrow[\text{A}]{-NH_3} \text{citrulline} \xrightarrow[\text{B}]{+P} \text{ornithine} \xrightarrow[\text{C}]{-CO_2} \text{putrescine}$$

Enzymes involved in the sequence of the reaction are:
　A—arginine deiminase
　B—ornithine transcarbamylase
　C—ornithine decarboxylase

It was shown that when *S. faecalis* ATCC 8043 cells were frozen prior to entrapment into acrylamide, citrulline alone was the major product formed ($>$ 90%). The microbial catalyst lost enzyme systems B and C; coupled with the observation that exogenous citrulline was not metabolized by intact immobilized cells, suggested that arginine could pass through the cell membrane, where cellular arginine deiminase transformed it to citrulline, but the citrulline remained inside the microorganism because the membrane acted as a barrier for its diffusion. Another explanation could be that accumulating citrulline inhibited the ornithine transcarboxylase enzyme, resulting in higher citrulline yield.

Thus, high product yield can be obtained if:
1.　A microbe with inhibited or no B and C enzyme systems can be immobilized, or
2.　The support matrix inhibits unwanted enzymes, or
3.　The support matrix affects cell wall/membrane permeability.

Pseudomonas putida　ATCC 4359 exhibited higher L-arginine deiminase activity and no ornithine transcarbamylase activity compared with other microorganisms. The cells were immobilized by entrapment into acrylamide gel and packed into a column (1.5 $\times$ 17cm). Transformation by the arginine deiminase enzyme of entrapped cells was determined as 56% that of free cells when the substrate, 0.5 M L-arginine hydrochloride (pH = 6.0), was passed through the column at 37°C. The low activity may be a manifestation of acrylamide toxicity or a substrate/product diffusion barrier caused by the lattice.

Several differences between immobilized and free cell properties were observed:
1.　Immobilized cells could be reused and exposed to a new substrate, whereas free cells could not.
2.　The relative activity of the immobilized cell system rose with increasing temperature: Maximum relative activity of L-citrulline formation (140%) by the microbial catalyst occurred at 55°C. For the free cell system, maximum relative activity (100%) was observed at 37°C and increasing the

temperature to 55°C resulted in drastic reduction of relative activity to only 20%.

3. After prolonged heat treatment, (60°C, 1–3 hour intervals), immobilized cells showed more heat stability than free cells.

4. Immobilized as well as free *P. putida* cells exhibited maximum L-citrulline formation in the pH range of 5.5–6.0. Immobilization did not change the pH optimum for L-citrulline production.

Effect of Substrate Flow Rate on L-citrulline Formation

0.05M L-arginine hydrochloride (pH = 6.0, 37°C) was passed through a packed column at various flow rates (SV). Moderately increasing the specific velocity from 0.16SV to 0.26SV raised substrate to product conversion to 100%. Further increases in specific velocity (0.39–0.49SV) lowered substrate transformation to 86% at 0.49SV.

Operational Stability of the Immobilized Cell Column

Half-life of the *P. putida* immobilized cell column was 140 days at 37°C and at a substrate flow rate of 0.19SV. Column half-life decreased rapidly with a temperature increase to 50°C.

The column effluent was concentrated by vacuum and dissolved in methanol; pure citrulline was recovered with a yield of 96%.

This process has been in industrial use since 1974.

Literature

Amin, G. *et al.*, *Eur. J. Appl. Micro.*, **14**, 59 (1982)
Kolot, F., *Dev. in Indst. Micro.*, **21**, 295 (1980)
Krouwell, P., *Biotech. Letters*, **2**, 5, 241 (1980)
Jack, T. *et al.*, *Adv. in Biochem. Eng.*, **5**, 126 (1977)

9

Synthesis of Organic Acids

Chapter 9 discusses production of industrially important organic acids by various microbial catalysts. Many of the systems described (aspartic, malic, uronic acids, 6-APA) are already being used in large scale applications. As in the previous chapter, for each process, the most active microbial strain, method of attachment, half-life, thermostability, as well as advantages and limitations are discussed.

9.0 Glutamic Acid Production

Monosodium glutamate (sodium salt of glutamic acid, MSG) is used therapeutically and as a flavor enhancing agent. Japan is the major manufacturer, producing 100,000 tons per year with *Corynebacterium glutamicus*, *Brevibacterium flavum*, *Brevibacterium divaricatum*, etc. strains. Microbial synthesis from glucose is a multi step enzymatic reaction requiring co-factors; the reaction sequence is as follows:

glucose→pyruvate→citrate→α-ketoglutarate→glutamic acid

Besides glucose, microbes can use acetic acid, ethanol, propylene-glycol, and paraffins as carbon sources.

Numerous research papers describe glutamic acid production by microbes isolated from different sources and by specifically deregulated auxotrophs; the effects of carbon and nitrogen sources, various biotin concentrations, aeration and agitation, as well as evaluation of the batchwise free cell system in a 50,000 liter fermentor are presented. The present section will review a less publicized though very competitive approach: application of living immobilized microbial cells for glutamic acid synthesis.

Corynebacterium glutamicus MB-1328 cells (biotin auxotroph) were immobilized by entrapment into two types of acrylamide gel, differing only in bisacrylamide content (gel A contained 9.2 gr of bisacrylamide, gel B contained 1.4 gr). The catalysts formed (20 gr each) were placed into identical fermentation media containing glucose (12.8%) and salts (pH = 6.8–7.2). Transformations was then performed aerobically with stirring (200 RPM) at 30°C.

After 144 hours of fermentation, the gel entrapped cells were removed from the reactors, washed, and reused. The amounts of glutamic acid formed continued to be monitored. Production by freshly prepared cells crosslinked into gel A increased linearly during 24–72 hours of operation, reaching a peak at 96–120 hr and accumulating 12–13 gr of glutamic acid per liter of medium. Reused cells entrapped into gel A exhibited only a slightly lower yield (10.5–11.0 gr/liter). A fresh catalyst prepared by cells entrapped into gel B synthesized nearly the same amount of glutamic acid as cells newly entrapped into gel A. Reusing gel B, however, resulted in lower product (5.0–7.0 gr/liter) formation compared with gel A. Although the cells immobilized into both supports remained stable after three months of storage at 4°C, gel A always exhibited higher activity than gel B (12.0–13.0 gr/liter vs 5.0–7.0 gr/liter, respectively). This result suggests that lower bisacrylamide content influences the support structure in a way that forms a barrier for the substrate.

The authors did not adapt their system to either column or fluidized bed reactor operation. By excluding biotin from the fermentation medium, they prevented growth of the biotin auxotroph inside the matrix, which probably reduced product yield. Nevertheless, the data clearly illustrate the feasibility of using gel-entrapped cells for continuous production of glutamic acid.

Constantinides harvested and entrapped an industrial strain of *Brevibacterium flavum* into reconstituted (desalted) collagen membranes when the cells were at their peak activity (after 10–12 hr of incubation). The membranes were used for glutamic acid production, utilizing glucose (3.6%) and urea (0.7%) as carbon and nitrogen sources. Substrate transformation was carried out in shake flasks at low agitation (25–30 strokes per minute, 32°C) and in a continuous recycle water jacketed column reactor (10 × 1.5 inches, with 1.0–1.5 gr of catalyst). An agitated and aerated medium was continuously recycled through the reactor for 24 hr. The used medium was then replaced, providing continuous operation.

The rate of product formation by the microbial catalyst was compared with that of free cells. Maximum glutamic acid production (9.0 gr/l) in the free cell system occurred at 65 hours of fermentation, with an agitation rate of 600RPM and an aeration rate of 1 vvm. Initial membrane activity in the immobilized column reactor approached 100%, but was only 40.9% for shake flasks. This observation indicates that although oxygen is necessary for product synthesis, too much causes a shift to succinic acid formation. It also signifies that oxygen transfer inside the column reactor was better than in shake flasks.

A significant effect of tanning conditions on catalyst stability was noticed. Tanning conditions are the time of collagen membrane exposure to glutaraldehyde and the concentration of the chemical used. With no treatment, half-life of the support was only three days, but tanning significantly improved collagen membrane stability. At a 0.5% glutaraldehyde concentration and a tanning time of one minute, half-life of the membrane increased to 9 days. However, due to its bacteriocidal nature, glutaraldehyde treatment reduced the initial relative activity of the cells to 75–80%.

Venkatasubramanian described the comparative economics of the conventional and immobilized cell processes for MSG production. Several assumptions were made:

1. Both plants utilize the same carbon source (15% dextrose), produce equal amounts of glutamic acid (5%), and have the same percent of product recovery (90%).
2. A fermentation plant includes ten fermentors, each with a 30,000 gallon capacity.
3. The immobilized cell reactor produces 0.05 pounds of glutamic acid per hour per pound of cells, and its operational half-life is three months.

Due to high fermentor cost, the fermentation plant required a larger total fixed capital investment (21.0 M) compared with the immobilized cell plant (12.9 M). The overall production cost for the immobilized cell plant (12.5 M) was greater than for the conventional plant (10.5 M) due to quarterly replacement of the immobilized cell catalyst. Based on the lower investment required for the immobilized cell process, the later provided 50% of return on investment compared to 36% for the conventional process. The economic advantages of the immobilized cell plant could be further improved by increasing the half-life and product yield of the cell catalyst.

Literature

Constantinides, A., *Biotech. Bioengin.*, **23**, 899 (1981)
Yamada, K., *Biotech. Bioeng.*, **19**, 1563 (1977)

9.1 6-aminopenicillanic Acid Synthesis

6-aminopenicillanic acid (6-APA) is a starting material for manufacturing synthetic penicillins. 6-APA is commercially produced from penicillin G or V by the enzyme penicillin acylase, (penicillin aminohydralase, E.C.3.5.11), which catalyzes deacylation of the penicillin molecule.

penicillin   6-APA

The reaction can proceed as chemical or enzymatic conversion (using free or immobilized enzymes), or as a continuous process using immobilized microbial cells.

The disadvantage of chemical conversion is its multi-step nature, requiring energy-intensive low temperatures and special equipment.

For enzymatic conversion, penicillin acylase must first be extracted, purified and then used as a free or immobilized preparation. Enzyme immobilization is performed by different techniques (adsorption, entrapment, coupling), using various supports. However, the products of penicillin G hydrolysis (6-APA and phenylacetic acid) are competitive and noncompetitive inhibitors of penicillin acylase. In addition, the immobilized enzyme partially loses its activity during continuous operation.

As an alternative to acylation by free and immobilized enzymes, the immobilized cell approach was investigated.

E. coli ATCC 9637, known as a penicillin acylase producing strain was attached to acrylamide by entrapment. Gel particles (3 mm diameter) were packed into a column (1.6 × 13 cm), and a penicillin G solution in borate-phosphate buffer (pH = 8.5) was pumped through the reactor at 30°C. During the course of 6-APA production, the cell catalyst was found to contain the enzyme penicillinase. Maximum product formation (12 μmole/ml) occurred after 3 hours of incubation with the substrate, but the rate of synthesis then declined due to the action of penicillinase, which decomposed 6-APA to 6-aminopenicilloic acid. Calculations revealed that 85–90% of 6-APA was formed during the 3 hour incubation period.

9.1.1 Enzymatic Properties of Immobilized Cells

Enzymatic properties of immobilized and free cells were compared; the following similarities and differences were observed:

1. The optimal pH for both preparations was 8.5.
2. The optimal temperature range for 6-APA formation was 37–40°C for both preparations.
3. The maximum 6-APA yield in the immobilized cell reactor was observed at a substrate (penicillin G in a borate-phosphate buffer, pH = 8.5) flow rate of SV = 0.12–0.24hr^{-1}.
4. Operational stability is one of the most important parameters of a cell catalyst. 6-APA production and catalyst stability were measured at various flow rates (SV = 0.12, 0.24, and 0.48). A decrease in the rate of product synthesis occurred after increasing SV from 0.12 to 0.48. The maximum formation of 6-APA (40 moles/ml) was observed at a space velocity of 0.12. At this flow rate, the catalyst remained fairly stable during 20 days of operation.

A solution of penicillin G in the borate-phosphate buffer was passed through a column at SV = 0.24. The effluent was adjusted to pH = 7.0 and concentrated in a vacuum. After two more pH adjustments and additional vacuum concentration, 6-APA was crystallized at 7°C with 6N (normal) HCl (pH = 4.3), giving a product yield of 7.8%.

The most recent results from the same laboratory (Chibata, et al., 1979, US Patent 4,138,292) described a further improved immobilized cell system. The product yield was increased to 90% and the half-life raised to 42 days. This system is in commercial use and is considered more advantageous than the continuous method using immobilized enzymes.

Literature

Zurkova, E., *Biotech. Bioeng.*, **25**, 9, 2231 (1983)
Chibata, I., US Patent 4,138,292 (1979)
Sato, T. *et al.*, *Eur. Jour. Appl. Micro.*, **2**, 153 (1979)
Hansher, J. *et al.*, US Patent 4,113,566 (1978)

9.2 Malic Acid Production

Malic acid is an essential compound of cell metabolism. It is used therapeutically for treatment of hyperammonemia, a hepatic malfunction. Conversion of fumaric acid to malic acid is accomplished by a one step enzymatic reaction:

$$\text{HOOC} - \text{CH=CH} - \text{COOH} \underset{\substack{\text{fumarase} \\ \text{(E.C.4.2.1.2)}}}{\overset{+H_2O}{\rightleftharpoons}} \text{HOOC} - \underset{2}{\text{CH}} - \underset{\underset{\text{OH}}{|}}{\text{CH}} - \text{COOH}$$

fumaric acid   malic acid

Fumarase catalyzes addition of a hydroxyl group to the double bond of fumaric acid. The reaction can be carried out as a batch process using free cells or free enzymes, and as a continuous process using immobilized enzymes. The most recent development in the field of malic acid production has been application of living immobilized cells. Industrial operation using immobilized microbes was introduced by the Japanese Tanabe Seiyaku Co. in 1974.

We will discuss and compare the properties of biocatalysts formed by entrapment into acrylamide and carrageenan gels. Changes in cell metabolism after immobilization as well as catalyst stability during continuous operation will also be presented.

9.2.1   Malic Acid Production by Cells Entrapped into Acrylamide Gel

A.   Screening for the Best Microbe

Several microorganisms (*Brevibacterium ammoniagenes* IAM 1645, *Corynebacterium equi* IAM 1038, *Escherichia coli* ATCC 11303, *Microbacterium flavum* IAM 1642, and *Proteus vulgaris* IFO 3045) immobilized by entrapment into acrylamide gel were tested for highest fumarase production. From the five strains examined, the *Brevibacterium ammoniagenes* IAM 1645 catalyst exhibited highest activity, 0.49m moles of L-malic acid/hr/gr cells. However, immobilized *Brevibacterium* cells transformed the substrate to a mixture of malic and succinic acids (95%:5%). Removing succinic acid from the final product presented difficulties, thus experiments were directed to suppress its synthesis. Acetone, cationic and anionic detergents, Triton x-100, bile acid, and bile extract were all tested at various concentrations as suppressors. Bile extract at the concentration of 0.2–0.3%, in the pH range of 5.0–7.0, and at the treatment temperature of 37°C most effectively inhibited succinic acid synthesis, reducing it to less than 0.2 mole % in the final product.

B.   Changes in Cell Properties after Immobilization

pH and temperature optima as well as half life of the immobilized cell matrix were compared with the free enzyme system. Maximum relative activity (100%) in both cases was observed at the pH of 7.0–7.5. The microbial catalyst had a similar temperature optimum (60°C) as did free enzymes, but was more stable during continuous operation. The half-life of the free enzyme system was only 6 days, while entrapped cells retained 90% of their activity during 10 days of operation. Immobilized cell stability was later improved significantly.

C. Continuous Column Operation: Effect of Temperature, Flow Rate, and Operational Stability

1M sodium fumarate (pH = 7.0) was pumped through an immobilized cell column (1.6 × 17.5 cm) at various space velocities (0.1–1.0) and temperatures (ranging from 29°C to 45°C). At 29°C, the maximum conversion rate of 70% occurred with SV = .1hr^{-1}, or about a 10 hr retention time. Increasing the temperature to 33°C caused the reaction to reach equilibrium faster. At 37°C, the reaction arrived at equilibrium when the substrate flow rates were below SV = 0.23, or above a 4.3 hr retention time. Under these conditions, 80% of substrate conversion was achieved.

Half-life of *B. ammoniagenes* cells entrapped into the acrylamide support was 53 days at 37°C. Product was recovered from the effluent by precipitation and filtration with a 70% yield.

To investigate the cause of a progressive decay in column activity, a sectional column reactor was used. Eight of the ten 116 ml sections were each packed with 70 gr of immobilized cells, while the top two contained only substrate (pH = 7.0) preincubated to 37°C, the temperature maintained in the water-jacketed column. When the substrate passed down the column, the activity in each section was monitored.

Only 20–40% of substrate conversion occurred in the top sections #1–4. Most malic acid (60–80%) was synthesized in the bottom sections #5–8. Increasing the medium flow rate resulted in lower product formation. The highest rate of substrate conversion (80%) was observed with the slowest flow rate (93 ml/hr). At 2400 ml/hr, product synthesis decreased sharply to only 18%. After 43 days of operation, 100% of the initial activity remained in the bottom sections, but only 80%, 83%, 89%, and 93% in sections 1–4.

The authors concluded that decay in enzymatic activity of the immobilized cell column may be contributed to:
1. Leakage of enzyme (or enzyme stabilizer) from the catalyst.
2. Contamination of the enzyme with poisonous substances derived from the substrate.
3. Irreversible denaturation of enzyme.
4. Toxicity of the support.

9.2.2 Malic Acid Production by Cells Entrapped into K-carrageenan Gel

A. Screening for the Best Microbe

A matrix significantly affects performance of the immobilized cell system, thus a change in support material should be followed by screening for the microbe most active with the chosen support.

To improve malic acid production, switching from the potentially toxic acrylamide gel to a carrageenan support was suggested. Of the two hundred and forty one strains screened, nine with high fumarase activity were tested on seven media of different compositions. As a result, four strains (*Brevibacterium ammoniagenes*, *Brevibacterium flavum*, *Proteus vulgaris*, and *Pseudomonas fluorescens*) were chosen for immobilization. The *B. flavum* catalyst formed by cell entrapment into the K-carrageenan gel exhibited the highest fumarase activity (8.28 m mole/hr/unit OD) and longest half-life (70 days). *B. ammoniagenes* microbes formed the second best catalyst: 5.80m moles/hr/unit OD, 60 day half-life.

Several factors which may affect fumarase enzyme activity are:
1. Thermal stability of the fumarase enzyme.
2. Conditions of gelation.
3. Carrageenan support and *B. flavum* cell concentrations.
4. The bile extract necessary to suppress succinic acid formation.

B. Changes in B. flavum *cell properties after Immobilization*

The properties of immobilized cells were investigated and compared with properties of free cells and free enzymes:
1. Immobilized cells showed a broader pH optimum (6.5–8.0) compared with free cells and free enzymes (7.0–7.5).
2. The optimum temperature for immobilized cells was 60°C, 10°C higher than for free cells and free enzymes.
3. The reaction reached equilibrium (850 μmoles of malic acid/ml) at flow rates below SV = 0.3 hr^{-1}.
4. Effect of the support matrix on relative productivities of immobilized *B. ammoniagenes* and *B. flavum* cells, and half-lives of the carriers formed are shown in Table 1.

Table 1   Productivities and Half-lives of Immobilized
B. ammoniagenes and B. flavum Strains

Immobilization Support and Strain Used	Fumarase Activity of Cells (μmoles·hr^{-1}·gr^{-1})	Half-life (days) at 37°C	Relative Productivity (%)
Polyacrylamide			
B. ammoniagenes	5800	53	100
B. flavum	6680	94	204
Carrageenan			
B. ammoniagenes	5800	75	142
B. flavum	9920	160	516

Activity of the catalyst obtained by *B. ammoniagenes* cell entrapment into the acrylamide gel was taken as 100%. With both microorganisms tested, relative productivity increased by using the carrageenan gel as the matrix, thus confirming acrylamide toxicity to the cells (noted previously by many authors). Under the same experimental conditions, the *B. flavum* catalyst formed by entrapment into carrageenan gel exhibited a 1.5 fold higher enzymatic activity compared with polyacrylamide entrapped cells. The 160 day half-life of the carrageenan catalyst was nearly two times longer compared with the polyacrylamide matrix (94 days).

The results clearly indicate the significant improvement in malic acid production obtained by switching support matrices. Assuming the use of a 1,000 liter column, 15.4 metric tons of L-malic acid can be produced during one month of operation at a substrate (1M sodium fumarate) flow rate of 200 liters/hr. L-malic acid can be separated by acidifying the effluent and recovering the product as a calcium salt. Using this method, 70% of the theoretical yield was obtained.

Literature

Wada, M. *et al.*, *Eur. J. Appl. Micro. & Biotech.*, **11**, 67 (1981)
Siess, M., Davies, C., *Eur. J. Appl. Micro. & Biotech.*, **12**, 10 (1981)
Takata, I. *et al.*, *Enzyme and Micr. Tech.*, **2**, 30 (1980)
Takata, I. *et al.*, *Eur. J. Appl. Micro. & Biotech.*, **7**, 161 (1979)

9.3   Citric Acid Synthesis

$$COOH - CH_2 - \overset{\overset{\displaystyle COOH}{|}}{\underset{\underset{\displaystyle OH}{|}}{C}} - CH_2 - COOH$$

Citric acid is used in foods, medicines, and by the chemical industry. It can be produced by various *A. niger* strains in surface or submerged fermentation, utilizing different carbon and nitrogen sources. The reaction is a multi-step enzymatic process, still carried out industrially as a batchwise operation. A potentially better alternative is performing continuous citric acid synthesis using living immobilized cells.

A. niger mycellium (72–96 hrs old), immobilized by entrapment into a collagen membrane, was used for product formation in a continuous flow reactor. Investigating the effect of substrate flow speed on the reaction rate revealed a linear relationship at medium velocities ranging from 50 to 150 ml/min. The function leveled off at 235 ml/min, giving a maximum product yield of 12 gr/gram catalyst/hr.

It was also observed that citric acid production depended on the thickness of the collagen membrane. Increasing its thickness from 7 to 20 mils reduced product yield, suggesting a substrate/product diffusion limitation. With the optimal membrane thickness of 7 mils, relative activity in the continuous immobilized cell reactor was two times higher than in shake flasks containing the same amount of catalyst. Raising the dissolved oxygen concentration in the medium to 80–90% of saturation increased the product yield drastically compared with shake flask experiments. This result emphasizes the importance of oxygen delivery to the attached microbes for effective column operation. Maximum relative specific productivity of this system was only 48.4% compared with free cells; the half-life of the catalyst was 138 hrs.

Deromatographic analysis of products formed by the immobilized cell reactor revealed the presence of isocitric (15–20%), oxalic, and traces of gluconic acids. The amounts of other organic acids formed by free cells during citric acid fermentation varied with the strain, carbon source, pH, aeration, etc. By properly matching the above parameters, the quantities of unwanted products synthesized by the immobilized cell catalyst can be reduced.

Literature

Horitsu, H. *et al.*, *Appl. Micro. & Biotech.*, **22**, 1, 8 (1985)
Eikmeier, H., *Appl. Micro. & Biotech.*, **20**, 6, 365 (1984)

9.4   Uroconic Acid Production

Uroconic acid is widely used in cosmetics and as a sunscreening agent. L-histidine can be converted to uroconic acid by the enzyme L-hystidine ammonium lyase:

A 0.25 M solution of histidine (pH = 9.0, 37°C) was passed through a packed bed reactor containing cells of *Achromobacter liquidum* IAM 1667 immobilized into acrylamide gel. Pure uroconic acid was produced and crystallized with a high yield. The half life of this system was 180 days.

Literature

Mosbach, K., *Ann. NY Acad. Sci.*, **434**, 239 (1984)
Jack, T., *Biotech. & Bioeng.*, **19**, 631–648 (1977)
Kan, J., Shuler, M., *Biotech. & Bioeng.*, **20**, 217–230 (1978)

10

Small Scale Applications

This short chapter reviews several applications of immobilized microbes that are presently being investigated. Continued research of these topics may make the processes economically feasible.

10.0 L-sorbosone Production

Sorbosone, a potential intermediate for the synthesis of vitamin C, can be produced from sorbose by microbial sorbose dehydrogenase:

The *Gluconobacter melanogenus* IFO 3293 strain was selected for highest enzyme activity and immobilized by entrapment into polyacrylamide gel. This method of attachment resulted in a loss of 32–48% of the free cell enzyme activity. Sorbosone production by the microbial catalyst increased during the first 40 hours of incubation, reached 1.3 mg/hour and then rapidly decreased to 0.1 mg/hour at 100 hours. Product formation in the column reactors was limited by a deficiency of dissolved oxygen, but aeration with pure O_2 caused rapid loss of activity. There were no shifts in pH and temperature optima compared with free cells; however, immobilized cells did show improved thermal stability. Sorbose dehydrogenase inactivation was prevented by a high concentration (20%) of sorbose, leading to the conclusion that sorbose is necessary for stabilization of proteins involved in the conversion. In the presence of 20%

sorbose, 310 mg of entrapped cells accumulated 450 mg of sorbosone during 360 hours of incubation.

Literature

Martin, C., Perlman D., *Biotech. and Bioeng.*, **18**, 217–237 (1976)

10.1 Coenzyme A Production

A multi-step enzymatic reaction was carried out by cells of *Brevibacterium ammoniagenes* IFO 12071 immobilized by entrapment into polyacrylamide gel. The substrate mixture contained sodium pantothenate (0.5 mM), cysteine (1.0 mM), ATP (1.5 mM), $MgSO_4$ (1.0mM), and buffer (pH = 7.5, 15mM). The catalyst was incubated in the substrate for 8 hours at 37°C, filtered, and again allowed to react with a fresh substrate mixture. The operation was repeated four times, yielding a total of 153 mg of CoA. Immobilized cells exhibited greater heat and pH stabilities compared with free cells, and had a high storage stability (45 days at 0°C). CoA production was also studied in a continuous column reactor (0.9 × 30 cm). The rate of synthesis declined about 50% after 5 days of continuous operation.

Literature

Shimisu, S., Marioke, H., Tani, Y., Ogata, K., *Journal of Fermentation Technology,* **53**, 2, 77–83 (1975)

10.2 Dihydroacetone Production

Dihydroacetone is an important organic compound widely used in medicines and pharmaceutics. Conversion of glycerol to dihydroacetone is a one step enzymatic reaction catalyzed by the enzyme glycerol oxidase:

$$
\begin{array}{ccc}
\text{H} & & \text{H} \\
| & & | \\
\text{H} - \text{C} - \text{OH} & \xrightarrow[\text{oxidase}]{\text{glycerol}} & \text{H} - \text{C} - \text{OH} \\
| & & | \\
\text{H} - \text{C} - \text{OH} & & \text{C} = \text{O} \\
| & & | \\
\text{H} - \text{C} - \text{OH} & & \text{H} - \text{C} - \text{OH} \\
| & & | \\
\text{H} & & \text{H} \\
\text{glycerol} & & \text{dihydroacetone}
\end{array}
$$

The conversion is a strictly aerobic process and can be carried out by *Acetobacter suboxydans* ATCC 621, *A. xylinum* A-9, *Gluconobacter melanoge-*

nus IFO 3293, *C. melanogenus* IFO 3294, and *Gluconobacter oxydans*, immobilized by entrapment into acrylamide, K-carrageenan, or Ca-alginate gels. The K-carrageenan gel, an "ideal support" for numerous one step as well as multistep enzymatic reactions, was inadequate for the glycerol-dihydroacetone transformation due to inhibition of the glycerol oxidase enzyme by metal ions necessary to maintain carrageenan gel strength.

Entrapped cells usually exhibited significantly lower glycerol-oxidase activity compared with free cells (18–60%), which may have resulted from matrix toxicity to the cells, a barrier to oxygen diffusion, or a combination of both. In fact, when the gel preparation was milled with a mortar, its activity increased to 60% of the free cell dihydroacetone production, suggesting that oxygen diffusion was the major factor limiting product formation.

10.2.1 Enzymatic Properties of Immobilized Cells

The enzymatic properties of free and immobilized cells of *A. xylinum* were compared:

1. Free cells exhibited a nearly symmetrical activity vs pH curve, which peaked at pH = 5.5, whereas immobilized microbes showed a broad pH optimum, with most activity occurring in the pH range of 4.0–5.5.

2. The optimal temperature for both immobilized and free cells was 35°C, but thermal stability of entrapped cells was greater at elevated temperatures (100% of initial activity at 47°C after 20 min.) than the stability of free cells (20% of initial activity at 47°C after 20 min.).

3. Raising the substrate concentration from 0.3M to 2M did not affect the relative activity of free cells. By contrast, the optimal glycerol concentration for immobilized microbes was 0.35–0.5M, further increases in which (0.5–1.5M) caused declining catalyst activity. Very similar K_m values were recorded for glycerol in free and immobilized cell systems (6.3×10^{-2}M and 6.5×10^{-2}M, respectively).

Although the microbial catalyst could be reused, its glycerol oxydase activity decreased significantly (up to 50%) compared with initial activity. No attempts were made to reactivate cells entrapped into the gel matrix, but since oxygen was found to be a limiting factor in dihydroacetone formation, various attempts were made to increase its availability to the entrapped cells. Transformation by the catalyst was performed under two conditions, using oxygen enriched air, and using hydrogen peroxide as the oxygen source.

By supplying oxygen enriched air, glycerol oxidase activity of acrylamide entrapped cells increased by 60% compared with free cells (from 2.81 grams of DHA/gr biomass/hr to 5.95 grams of DHA/gr biomass/hr). Using the second oxygen source, the *Gluconobacter oxydans* ATCC 621 strain also exhibited high catalase and glycerol oxidase activity. Microbes decomposed hydrogen peroxide inside the matrix and utilized the liberated oxygen for product formation.

The effect of hydrogen peroxide on catalyst activity was investigated with *G. oxydans* ATCC 621 cells immobilized by entrapment into an alginate gel. The beads formed were incubated at different concentrations of H_2O_2 (5–100mM). At the low hydrogen peroxide concentration of 20mM, 100% of the initial relative activity still remained. Increasing the peroxide concentration from 10 to 100mM reduced the time needed for dihydroacetone formation from twelve hours to only two. High H_2O_2 concentrations (50–100mM), however, drastically diminished relative activity of the catalyst.

10.2.2   Continuous Production of Dihydroacetone

Continuous production of dihydroacetone by Ca-alginate entrapped cells was performed in a column type reactor (bed volume 40 ml). The column was fed with 0.1M glycerol and 5.0mM $CaCl_2$ in a succinic buffer (pH = 5.0) at a substrate flow rate of 1.5 ml/minute. During the first 20 hours, an air saturated medium was pumped through the reactor. For the next 20 hours, the medium was saturated with oxygen, and after a total of 40 hours, it was supplemented with hydrogen peroxide at a concentration of 25mM.

Throughout the first time period, DHA productivity remained zero, and increased only slightly after the medium was saturated with oxygen (0.01 gr/hr). Adding the hydrogen peroxide (25mM) raised DHA synthesis to 0.19 gr/hr, but the half-life of the system fell to only 15 hours. The relatively high concentration of peroxide in the medium caused rapid deterioration of column activity.

Using hydrogen peroxide in glycerol transformation is only one of the means for supplying oxygen to the catalyst. This method merely emphasizes the severity of the oxygen delivery problem when applying immobilized cells in a strictly aerobic process. Controlled oxygen-peroxide feeding at a low but sufficient level can only minimize its adverse effect, but cannot eliminate it.

Low activity of the immobilized cell catalyst after entrapment into acrylamide gel was another significant problem which may reflect matrix toxicity, leakage of pyridine nucleotides (a cofactor) from the immobilized cells, or the inability of entrapped cells to completely regenerate the oxidized pyridine nucleotide from the reduced form. Attempts to improve cofactor stability by adding nicotinamide adenine dinucleotide or nicotinamide adenine dinucleotide phosphate to the reaction mixture were unsuccessful. However, cofactor regeneration by cell reincubation in a nutrient medium was not considered.

Until solutions for oxygen delivery and cofactor regeneration are found, glycerol-dihydroacetone conversion by the immobilized cell system offers no advantage compared to free cell transformation.

Literature

Holst, O. *et al.*, *Eur. J. Appl. Micro. & Biotech.*, **14**, 64 (1982)
Makhotkina, T. *et al.*, *Appl. Biochem. and Micro.*, **17**, 1, 102 (1981)
Yamada, S. *et al.*, *J. Ferm. Tech.*, **57**, 215 (1979)
Nabe, K. *et al.*, *Appl. Envir. Micro.*, **38**, 6, 1056 (1979)

10.3   Immobilized Cells for Analytical Purposes: Microbial Sensors

Microbial sensors are widely used for monitoring metabolites in continuous flow analyses. Methods based on application of biocatalysts are highly sensitive, substrate specific, reproducible, and less time consuming compared with traditional techniques. The concentration of a metabolite in a microbial sensor is determined by measuring cell respiratory activity via an attached oxygen (or other) probe. The following is an example of a sensor for determining the concentration of ammonia in wastewater.

Nitrifying bacterium *Nitrosomonas europaea* utilizes ammonia as its sole source of respiratory energy via the following reaction:

$$2NH_3 + 3O_2 \rightarrow 2HNO_3 + 2H_2O$$

The bacterium was immobilized by filtration through a porous acetylcellulose membrane (type HA, $0.45\mu m$ pore size, 25mm diameter, $150\mu m$ thickness), which was carefully attached to the Teflon membrane of the oxygen probe (so that the cells were trapped between the two membranes), covered with a nylon net, and fastened with a rubber ring. The microbial sensor was placed into a water jacketed flow column (constant temperature), and current from the oxygen probe was recorded as a buffer solution saturated with dissolved O_2 passed through the reactor.

When a sample solution containing ammonia was injected into the system, it permeated through the porous acetylacetate membrane and was assimilated by the immobilized bacteria. Consumption of O_2 by the microbes lowered the dissolved oxygen concentration around the membrane in direct proportion to $[NH_3]$ of the sample solution. As a result, current from the probe decreased markedly with time until a steady state was reached, indicating that consumption of O_2 and its diffusion from the sample solution to the membrane were in equilibrium.

Literature

Nakamura, N., Murayama, K., Kinoshita, T., *Anal. Biochem.*, **152**, 2, 386 (1986)
Owen, V., *Ann. Clin. Biochem.*, **22** (pt 6), 559 (1985)

INDEX OF MICROBES

INDEX